The Earth Policy Reader

Other Norton Books by Lester R. Brown

Eco-Economy: Building an Economy for the Earth

State of the World 1984 through *2001*
annual, with others

Vital Signs 1992 through *2001*
annual, with others

Beyond Malthus
with Gary Gardner and Brian Halweil

The World Watch Reader 1998
editor with Ed Ayres

Tough Choices

Who Will Feed China?

Full House
with Hal Kane

The World Watch Reader 1991
editor

Saving the Planet
with Christopher Flavin and Sandra Postel

Building a Sustainable Society

Running on Empty
with Colin Norman and Christopher Flavin

The Twenty-Ninth Day

In the Human Interest

Earth Policy Institute® is a nonprofit environmental research organization dedicated to providing a vision of an eco-economy and a roadmap on how to get from here to there. It seeks to reach a global constituency through its publications and the media. Its primary publications are *Eco-Economy: Building an Economy for the Earth*, *The Earth Policy Reader*, and a series of four-page Eco-Economy Updates that assess progress, or the lack thereof, in building an eco-economy. All of these can be downloaded at no charge from the EPI Web site.

Web site: <www.earth-policy.org>

THE EARTH POLICY READER

Lester R. Brown
Janet Larsen
Bernie Fischlowitz-Roberts

EARTH POLICY INSTITUTE®

W · W · NORTON & COMPANY
NEW YORK LONDON

Printed in the United States of America

First Edition

The text of this book is composed in Sabon. Composition by Elizabeth Doherty; manufacturing by The Maple-Vail Book Manufacturing Group.

ISBN 0-393-32406-0

W. W. Norton & Company, Inc., 500 Fifth Avenue,
New York, N.Y. 10110
www.wwnorton.com

W. W. Norton & Company, Ltd., Castle House, 75/76 Wells Street,
London W1T 3QT

1 2 3 4 5 6 7 8 9 0

This book is printed on recycled paper.

Contents

2. Eco-Economy Indicators: Twelve Trends to Track

3. Eco-Economy Updates

Energy and Climate

Population and Health

ACKNOWLEDGMENTS

Earth Policy Institute exists today because Roger and Vicki Sant of the Summit Foundation shared a vision of an eco-economy and provided an extraordinarily generous startup grant last year. The Sants were joined by several foundations, including the Farview, Richard and Rhoda Goldman, William and Flora Hewlett, W. Alton Jones, Shenandoah, Turner, and Winslow foundations and the U.N. Population Fund, which generously supported our effort. The enthusiasm of two foundation heads, Harriett Crosby of the Farview Foundation and Ted Turner of the Turner Foundation, went beyond providing grants as they purchased copies of *Eco-Economy* by the carton for their own special distributions.

As a startup organization, it is especially heartening to receive unsolicited individual contributions. We are particularly indebted to Robert Wallace and Raisa Scriabine for their very generous contribution. I would also like to thank Junko Edahiro, who serves as my interpreter in Japan, for her generous contribution, and for organizing a highly successful fundraiser in Japan. The other key donors include Douglas and Debra Baker, Maureen Hinkle, Hadan and Reah Janise Kauffman, Tamae Kobayashi, James McManis, Scott and Hella McVay,

Paul Myers and Welthy Soni, and Peter Seidel. We also thank the many other individuals who supported the Institute with contributions.

Production of this book was facilitated by an experienced team including three veterans—Linda Starke, Reah Janise Kauffman, and myself—who have produced many books together over a stretch of 16 years. Linda Starke, who has edited all the books I've written over the past 20 years, brought to the project both her editorial talent and deep knowledge of the issues. Her experience and insights were invaluable in coordinating production of this complexly structured book with three authors.

The other "veteran," Reah Janise Kauffman, Vice President of Earth Policy Institute, has worked with me for 16 years. Her understanding of the issues, of all of the phases of producing a book as well as running an organization, and of outreach, along with her enthusiasm and insight, are appreciated more than words can say. And in our small staff, where each of us wears several hats, Reah Janise also manages our worldwide publishing network, assists with fundraising and marketing, and manages the Web site.

Manager of Publications Sales Millicent Johnson handles our bookkeeping, makes sure all of our office needs are covered, and keeps us in good cheer. Her customer relations skills and knowledge of the intricacies of databases are invaluable. Elisha Triplett, Administrative Assistant, very ably manages our library, listserv, and news clips, and assists the research staff and occasionally me as well. Sadly we must soon bid Elisha farewell as she heads off to graduate school this fall.

And while it is not customary to thank coauthors, I would like to thank Janet Larsen, Research Associate, and Bernie Fischlowitz-Roberts, Staff Researcher, for their contributions to this volume. Janet helped me with research while also researching and writing her own pieces. Her leadership on this project has made my work much easier. When Bernie joined us last December he had hardly gotten his jacket off before he was assigned responsibility for making preliminary estimates of global wind power generating capacity for 2001. Since then he has worked on a wide range of climate and energy issues. The combined efforts of Janet and Bernie and their unflagging support contributed enormously to the finished product.

We were also fortunate to have Elizabeth Doherty, with whom we worked for four years at Worldwatch, doing the layout and design of the book. We benefited from both her professionalism and her sense of humor.

We are indebted to those who have reviewed various parts of this book. Toby Clark provided invaluable comments on Part 1, drawing on his long-standing interest in the interface between economics and ecology to offer numerous useful suggestions. Maureen Kuwano Hinkle drew on her 18 years of experience working on agriculture for the Audubon Society to provide useful comments on Parts 1 and 3. Reah Janise Kauffman made some structural suggestions that greatly improved the book. Others who reviewed various Eco-Economy Updates include Lisa Mastny and Randy Swisher.

Part 3 of *The Reader* is a compilation of previously released Eco-Economy Updates. Along this line, we would like to thank a unique group of people—those who voluntarily translate the Updates into other languages to be posted on Web sites and distributed on listservs. Li Kangmin translates them into Chinese <www.fish.net.cn>. UNESCO Catalunya translates them into Catalan <www.unescocat.org/llista.html>. Gianfranco Bologna translates them into Italian <www.wwf.it/worldwatch/cover.html>. Soki Oda translates them into Japanese <www.worldwatch-japan.org>, and Junko Edahiro sends out a Japanese translation via her listserv. In Brazil, Eduardo Athayde translates them into Portuguese <www.wwiuma.org.br> and distributes them on his listserv. Leif Ohlsson translates them into Swedish <www.smvk.se/Omvarldsbilder>.

We also thank those who supplied research information for the various components of the book. These include Farzaneh Bahar, Natalie Bailey, Adele Crispoldi, Fred Crook, Mike Ewall, Andrew Field, Robert Freling, Fred Gale, Ben Halpern, Nalin Kishor, Frank Langfitt, Wei Lui, Paul Maycock, Daniel Pauly, Gabe Petlin, Caroline Pollock, Farzaneh Roudi, Kai Schlegelmilch, Matthew Sones, Hamid Taravati, Sen Wang, Tao Wang, Christopher Ward, Timothy Whorf, and Sun Xiufang.

As we worked on the trends section, we were again reminded of how valuable the Worldwatch database is. *Vital Signs 2002* was especially helpful.

The team members at W.W. Norton & Company have been a pleasure to work with. This group includes Ingsu Liu, who patiently worked with us through various cover designs; Amy Cherry and Lucinda Bartley in the editorial department, who helped us keep to our deadlines; and the production team led by Andrew Marasia, who put the *Reader* on the fast track.

Lester R. Brown

PREFACE

The Earth Policy Institute, now just over a year old, is a new kind of research organization. It offers not only a vision of an environmentally sustainable economy—an eco-economy—but also frequent assessments of progress in realizing that vision. An early vision of an eco-economy was presented in *Eco-Economy: Building an Economy for the Earth*, published in November of 2001.

Since completing *Eco-Economy*, the Institute has released four-page Eco-Economy Updates every two weeks or so. These are included in Part 3 of this *Earth Policy Reader*.

One of Earth Policy's distinguishing features is its effort to reach a global constituency. To do this, it relies heavily on a worldwide network of publishers to translate and market its books. *Eco-Economy* is being published in 17 languages and 21 editions. The languages include Arabic, Catalan, Chinese, French, Indonesian, Italian, Japanese, Korean, Persian, Polish, Portuguese (in Brazil), Romanian, Russian, Spanish, Turkish, and Ukrainian. There are three English editions (US/Canada, UK/Commonwealth, and India), two Chinese editions (mainland China and Taiwan), and two Spanish editions (Spain and Latin America). Most of the publishers in this network will also be publishing *The Earth Policy Reader*. (For information on the various language publications, please refer to our Web site, <www.earth-policy.org>.)

Earth Policy relies heavily on the communications media, both print and electronic, to disseminate its research results. These include, for example, the big three global electronic networks—the BBC, Voice of America, and CNN—and the major-language wire services, ranging from the German Press Agency to Xinhua in Chinese.

Each Eco-Economy Update is automatically distributed to our worldwide press e-mail and fax lists, totaling some 1,200 editors and reporters. One of the highlights of publishing *Eco-Economy* was the 14 excerpts published in magazines, including *The Ecologist*, *The Futurist*, *The Humanist*, *Mother Earth News*, and *Solar Today*. *Luup*, a Finnish news magazine, did a seven-page cover story on *Eco-Economy*.

Perhaps the most dynamic component of the Institute's outreach program is the dissemination of information via the Internet. The Updates are distributed on the Institute's own listserv, which now contains nearly 7,000 addresses, including both individuals and organizations. We are delighted that many of them now distribute the Updates on their own listservs.

In addition to news coverage of the Updates, many newspapers and magazines print them in their entirety. Among the newspapers that regularly carry the Updates are *Yomiuri* in Japan, which with a circulation of 10.2 million is the world's largest newspaper; *The Hindu* in India, a national English language newspaper; and *Hamshahri*, a leading Iranian newspaper. The Korean Federated Environmental Movement publishes them in its monthly magazine, *Hamggesaneungil*.

There are also more than a dozen electronic news networks and Web sites that regularly carry or cover the

Updates, such as the Environment News Service and the UN Foundation Newswire.

The enthusiasm with which the Updates have been received led us to compile them in this book. Part 3, entitled Eco-Economy Updates, contains some 20 Updates. Several others have been used in Part 2, in the Eco-Economy Indicators section, a collection of 12 trends to track in assessing progress in building an eco-economy.

Part 1 of the book looks at the ecological deficits that the world is now facing and at their effect on the food and energy economies. It begins with an in-depth examination of the effect of ecological deficits on China, which is now being invaded by advancing deserts. It contains some of our latest research findings, including those from my May 2002 field trip in China, which included Inner Mongolia and Gansu, two provinces directly affected by expanding deserts.

The Earth Policy presence in China has benefited greatly from the efforts of Lin Zixin. Mr. Lin worked closely with the International Technology and Economics Institute, an arm of the State Council, which sponsored *Eco-Economy* and hosted my visit to China. He also assembled the skilled team of translators and supervised the production of the Chinese edition for the People's Publishing House.

We have been pleased with the response of political leaders to *Eco-Economy*. The Romanian edition, for example, was launched by President Ion Iliescu in late January 2002.

The corporate interest in *Eco-Economy* was reflected in an invitation to address the Business Council. *Worldlink*, the magazine of the World Economic Forum in Davos, published a five-page feature adapted from *Eco-Economy*. *U.S. News and World Report* devoted three pages to a story on the enormous potential of wind power, which was inspired by *Eco-Economy*.

Within the foundation community, a number of foundations distributed copies of *Eco-Economy* to their directors, including the Educational Foundation of America, the Rasmussen Foundation, the Summit Foundation, the Turner Foundation, the Wallace Genetic Foundation, and the Weeden Foundation. In addition, the Turner Foundation distributed nearly 1,000 copies to their grantees, seeing it as a way to develop a shared vision of an eco-economy. Environmental groups that are marketing *Eco-Economy* range from the American Wind Energy Association to Population Connection (formerly Zero Population Growth).

As part of our dissemination effort, all of our research products, including this book, are available on our Web site. Permission for reprinting or excerpting portions of the book can be obtained from Reah Janise Kauffman at <rjkauffman@earth-policy.org> or by fax or mail. If you would like to receive our Eco-Economy Updates as they are released, please sign up via our Web site, or send us a note by e-mail.

We welcome your input, including any thoughts or recent papers or articles you would like to share with us.

Lester R. Brown

Earth Policy Institute
1350 Connecticut Ave., NW, Suite 403
Washington, DC 20036

phone: 202.496.9290
fax: 202.496.9325
e-mail: epi@earth-policy.org
Web site: www.earth-policy.org

July 2002

The Earth Policy Reader

1

The Economic Costs of Ecological Deficits

As populations have multiplied and incomes have risen, demands on the natural support systems of many economies have become excessive, generating ecological deficits. The effects of these deficits are first seen at the local level as deforestation leads to fuelwood shortages, overplowing leads to falling crop yields, overgrazing leads to emaciated herds of cattle, or overpumping drops water tables and dries up wells.

At some point, these expanding deficits begin to reinforce each other, creating an ecological disaster of national proportions. This is now happening in China—where disappearing forests, deteriorating rangelands, eroding croplands, and falling water tables are converging to create a dust bowl of historic dimensions. China's sheer geographic size, the weight of its 1.3 billion people on the land, and the pace of its economic expansion put it on the frontline of the deteriorating relationship between the global economy and the earth's ecosystem.

Although China is not well prepared for it, it is now at war. It is not invading armies that are claiming its territory, but expanding deserts. Old deserts are advancing and new ones are forming, like guerrilla forces striking unexpectedly, forcing Beijing to fight on several fronts.

And China is losing the war. Not only are the deserts advancing, but their advance is gaining momentum, claiming an ever larger piece of territory each year. The flow of refugees has already begun, as villages in several provinces are overrun by sand dunes.

Mounting ecological deficits are taking an economic toll in other countries as well. Algeria, suffering from the same complex of deficits as China, is trying to convert the southern one fifth of its grainland to orchard crops in an effort to halt the advancing Sahara. On the Sahara's southern fringe, Nigeria is fighting a similar battle.

Among other things, these ecological deficits are producing refugees. Villages are being abandoned as aquifers are depleted in Iran and India. Kazakhstan has surrendered half of its cropland to the desert.

On another front in this war, if sea level rises by 1 meter during this century, which is now clearly a possibility, Bangladesh will lose half of its riceland, scores of other Asian countries will lose their rice-growing river deltas, and some island countries will become uninhabitable. Modern civilization, with its growing population, is being squeezed into an ever smaller area by expanding deserts and rising seas.

The first section of Part 1 of *The Earth Policy Reader* describes how ecological deficits are converging to expand deserts in China. The second section looks at the negative effect of ecological deficits on the food prospect and how to eliminate both soil and water deficits. The third section describes how nature's inability to fix carbon as fast as we release it is destabilizing climate. It then discusses how to restructure the energy economy and eliminate this deficit by reducing carbon emissions. And finally we discuss how to fix the market to achieve environmental sustainability.

Deserts Invading China

Lester R. Brown

On April 12, 2002, South Korea was engulfed by a huge dust storm from China that left residents of Seoul literally gasping for breath. Schools were closed, airline flights were cancelled, and clinics were overrun with patients who were having difficulty breathing.[1]

The health effect was pervasive. When the amount of particulate matter in the air—normally 70 micrograms of dust per cubic meter in Seoul—reaches 1,000, respiratory stress disables the elderly and those with impaired respiratory systems. At the 2,070 micrograms recorded in this particular dust storm, breathing was labored for the able-bodied as well as the infirm. Many people were afraid to venture outside.[2]

New York Times correspondent Howard French reports that these suffocating dust storms, once seen as a nuisance in Korea, are now considered an economic threat as they boost worker absenteeism, curb travel, reduce retail sales, and adversely affect dust-sensitive industries, such as semi-conductor manufacturing. The automobile maker Hyundai began to shrink-wrap cars destined for export as soon as they came off the assembly line lest they arrive in foreign markets saturated with dust. Both business and tourist travel are reduced when the country is besieged by these dust storms. Airline flight cancellations are increasingly common.[3]

Koreans have come to dread the arrival of what they now call "the fifth season"—the season of dust storms that occupies the months on the calendar once considered late winter and early spring. Japan also suffers from dust storms originating in China. Although not as directly exposed as Koreans are, the Japanese complain about yellow snow and the brown rain that streaks their windshields and windows. It is Korean and Japanese frustration with Chinese dust that led to the launch of a trilateral ministerial consultation between South Korea, China, and Japan in 1999.[4]

Occasionally even the United States is affected. In April 2001, a huge dust storm measuring 1,800 kilometers east-west and 1,200 kilometers north-south crossed the Pacific intact, blanketing the western United States from the Arizona border to Canada with dust. Atmospheric scientists in Boulder, Colorado, who sent a plane up to measure dust concentrations at every thousand feet detected dust up to 37,000 feet. In March 2002, another storm from China followed the jet stream east, crossing the western United States before dissipating over Colorado.[5]

Within China, the area affected is expanding as the number and size of the storms has increased in recent years. In late January 2002, an unusually early dust storm moved southward over Tibet, closing the airport in Lhasa for three days, disrupting tourism and other activities. In eastern China, dust storms reach the coastal populations as far south as Shanghai.[6]

While the dust storms can have severe effects in South Korea, they can be even more suffocating for the people of eastern China who are more directly affected. Early each year, residents of eastern cities, such as Beijing and

Tianjin, hunker down as the fifth season begins. Motorists have learned to drive with their lights on during the day as storms impair visibility, and residents routinely cover their faces with surgical masks, shawls, or handkerchiefs.[7]

The fifth season is not a pleasant one for those living in northern and eastern China. Those with respiratory illnesses are particularly burdened, as the breathing stresses intensify their illnesses. Apart from the difficulty breathing and the dust that stings the eyes, there is the constant effort to keep dust out of homes and to clear doorways and sidewalks of dust and sand.

As difficult as life may be for those living in the paths of the dust storms, the real price is paid by the pastoralists and farmers who live at their source. They are bearing the brunt of the dust and sand storms.

Although global media coverage of dust and sand storms in the more remote northern and western regions has been limited, enough time has now passed for the extent of damage from past storms to be measured and recorded in scientific papers. One of these reported on a dust and sand storm occurring on May 5, 1993, in the Hexi corridor of Gansu Province in China's northwest. This intense sand and dust storm reduced visibility to zero and the daytime sky was described as "dark as a winter night." The storm destroyed 170,000 hectares of standing crops, damaged 40,000 trees, killed 6,700 cattle and sheep, blew away 27,000 hectares of plastic greenhouses, injured 278 people and killed 49. Forty-two trains, both passenger and freight, were either cancelled, delayed, or simply parked to wait until the storm passed and the tracks were cleared of sand.[8]

A detailed record of the effect of a dust-sandstorm on April 5, 1998, in Alxa Prefecture in Inner Mongolia describes the damage from a storm that lasted for 12 hours. Some 10,600 hectares of crops were destroyed, including 330 hectares of wheat covered by shifting sand; 134 plastic greenhouses were damaged; 400 drinking wells were filled with sand; 130 hectares of fruit orchards were destroyed; 800 tons of hay and dry forage stored in open fields were simply carried away; 600 sheep sheds were damaged; 1,000 yurts were destroyed; and 7,000 sheep were killed. These accounts describe just two of the scores of sand and dust storms that have occurred in the last decade or so.[9]

Data are now becoming available on ecosystem decline in at least some locations in the more severely affected areas. In Alxa Prefecture, more than 3 million hectares of grazing land are degraded, of which 60 percent is seriously degraded. Fodder production in the region has decreased by 43 percent and the carrying capacity of the grazing land has declined by 46 percent. And perhaps most telling of all, the body weight of the average draft animal has been reduced by almost half, suggesting rather emaciated animals. The forested area in the region, which totaled 1.13 million hectares in the 1950s, has now shrunk to 530,000 hectares—most of which is in an unhealthy state. This sort of ecosystem deterioration can be found in numerous prefectures and counties in northern and western China.[10]

Advancing Deserts Gaining Momentum

Desertification in China is the product of excessive human and livestock pressure on the land in a country whose population will reach 1.3 billion next year—nearly as large as the world population of 1.5 billion when the twentieth century began. Under this demographic pressure, China is running up ecological deficits on many fronts: overgrazing its rangelands, overplowing its land, overcutting its forests, and overpumping its aquifers.

With little vegetation left in parts of northern and western China, the strong winds of late winter and early spring can generate a dust storm that removes literally millions of tons of topsoil in a single day, soil that can take centuries to replace.[11]

Desertification is the degradation of land associated with the loss of topsoil that follows loss of vegetation. The fine particles in soil exposed to the wind are the first to blow away, creating dust storms. Once the fine particles are gone, leaving only the course particles or sand, then sand storms occur. Dust storms can cover vast areas and travel great distances, whereas sand storms are more localized.

This conversion of productive land into wasteland is not new in China. Historical accounts refer to dust storms some 27 centuries ago. What is new is their frequency and scale. Dust storms in 2001 and 2002 were more numerous, larger, and much more disruptive than in previous years.[12]

The process of desertification itself directly affects 40 percent of China's landmass, including Sinkiang Province and Tibet in the far west and Qinghai, Gansu, Ningxia, and Inner Mongolia Provinces in the north-central region. Although desertification is concentrated in these six provinces, it is now spreading into Sichuan, Shaanxi, Shanxi, and Hebei Provinces as well.[13]

Scientists at the Cold and Arid Regions Environmental and Engineering Research Institute (CAREERI) in Lanzhou, the world's premier desertification research institute, believe that desertification is one of the most serious environmental problems. They have charted the nationwide growth of the area converted to desert over the last half-century. Each decade, the area has increased. Wang Tao, Deputy Director of CAREERI, reports that during the 1950s, 1960s, and 1970s, the average rate of desert spread was 1,560 square kilometers per year. During the 1980s, this expanded to 2,100 square kilometers a year—an increase of 35 percent. During the 1990s, the area converting to desert rose to 2,460 square kilometers per year—a further jump of 17 percent.[14]

The scientists in Lanzhou are hoping that the area turning to desert will be reduced during this first decade of the new century. But with the dramatic jump in the number, severity, and size of dust storms in 2000–02, the growth of the deserts appears to be accelerating.

China's Environmental Protection Agency reports that the Gobi Desert expanded by 52,400 square kilometers (20,240 square miles) from 1994 to 1999, an area nearly half the size of Pennsylvania. This figure does not include the spread of the large Taklimakan Desert, which is further west; the five smaller deserts in Inner Mongolia; or the many new deserts that are beginning to form. With the advancing Gobi now only 150 miles from Beijing, China's leaders are beginning to realize the gravity of the situation.[15]

Desertification is typically concentrated on the fringes of existing desert, simply because these are the areas of marginal rainfall with the least vegetation. Of even more concern, however, are the new desert areas, replete with sand dunes, forming spontaneously in so many communities in northwestern and northern China. Localized sand dunes forming within 80 kilometers of Beijing are alarming government officials.[16]

Data for major dust storms as compiled by the China Meteorological Agency also indicate that desertification is accelerating. After increasing from 5 in the 1950s to 14 during the 1980s, the number leapt to 23 in the 1990s. The new decade has begun with more than 20 major dust storms in 2000 and 2001 alone. If this annual rate continues throughout the decade, the total will jump to 100—a fourfold increase over the last decade. (See Table 1–1.)[17]

In addition to the land already converted to desert, 900,000 square kilometers (347,000 square miles) of the Chinese landscape show a clear "tendency toward desertification," according to Qu Geping, formerly Minister of Environment and now Chairman of the Environment and Resources Committee of the National People's Congress. This area of 90 million hectares, which consists mostly of rangeland but includes some cropland as well, is roughly equal to the area planted to grain in China.[18]

There is a tendency in viewing desert expansion to think of it in linear terms, but it may not in fact be linear beyond a certain point. For example, as livestock numbers increase and the forage supply deteriorates as a result of overgrazing, the situation may reach a point where the degradation accelerates, leading to rapid,

Table 1–1. *Number of Major Dust Storms in China, by Decade, 1950–99, with Projection to 2009*

Decade	Number
1950–59	5
1960–69	8
1970–79	13
1980–89	14
1990–99	23
2000–09	100[1]

[1]Preliminary estimate for decade based on more than 20 storms during 2000 and 2001.

Source: China Meteorological Administration, cited in "Grapes of Wrath in Inner Mongolia," report from the U.S. Embassy in Beijing, May 2001.

wholesale destruction. Once human and livestock populations start retreating from the advancing desert, the pressures from concentrating human and livestock populations on the desert fringe can become even greater. This, too, can accelerate ecosystem collapse. There is some evidence that this is now happening in China.

From Ecological Deficits to Dust Bowl

As noted earlier, several ecological deficits are converging in China to create a dust bowl on a scale never before witnessed. In its effort to remain self-sufficient in grain, China has tried to avoid any shrinkage in overall cultivated area. As industrialization has claimed cropland in the coastal provinces, the national policy has offset these losses with the cultivation of land elsewhere. Thus the cultivated area in some northern provinces expanded during the 1990s. In Inner Mongolia, for example, it grew by an astonishing 22 percent between 1987 and 1996.[19]

China's expanding demand for food has pushed agriculture onto marginal land in the northwestern provinces, much of it land too dry to sustain cultivation. As a result, the soil is blowing away and the land is losing its productivity. Eventually the unproductive cropland is abandoned. Traveling by train through northern and western China in May 2002, I saw many such plots of abandoned land.

Overgrazing may be even more damaging. China's herds of cattle and flocks of sheep and goats have outgrown the carrying capacity of rangelands, leading to a forage deficit. After the 1978 economic reforms, when China shifted to a market economy, the government lost control of livestock numbers. As a result, the livestock population has grown by leaps and bounds, far exceeding that of the United States, a country with a comparable grazing capacity. While the United States has 97 million cattle, China has 128 million. The United States has 8 million sheep and goats; China has a staggering 290 million. The sheep and goats that range across the land are simply denuding western and northern China, a vast grazing commons. In China, as in many other countries with common grazing areas, there is no administrative mechanism for limiting livestock populations to the sustainable yield of rangelands.[20]

A report by a U.S. embassy official in May 2001 after a visit to Xilingol Prefecture in Inner Mongolia notes that official data classify 97 percent of the prefecture's 200,000 square kilometers as grassland, but a simple visual survey indicates that a third of the terrain appears to be desert. A similar survey by an aid official for another prefecture in Inner Mongolia indicates that half of the land is now desert. The embassy report on Xilingol describes the livestock population in the prefecture jumping from 2 million as recently as 1977 to 18 million in 2000. (See Table 1–2.)[21]

The report notes that whereas the traditional nomadic herders kept a mix of horses, cattle, sheep, and goats, today's herds consist overwhelmingly of sheep and goats. And *People's Daily* reports that the yield of forage from Inner Mongolia's rangelands has declined by at least 30 percent, and perhaps as much as 70 percent, over the last half-century. A Chinese scientist doing grassland research in Xilingol estimates that if recent trends of desertification continue, Xilingol will be uninhabitable in 15 years.[22]

As China's population has grown, so too has the demand for fuelwood and lumber. Throughout most of the country, this demand now exceeds the sustainable growth of trees and shrubs. As a result, vegetation has disappeared in many areas, leaving little to hold the soil when the wind blows or when it rains.[23]

Deforestation of the southern provinces may also be

Table 1–2. *Livestock Population of Xilingol Prefecture, Inner Mongolia, 1977–2000*

Year	Number of Livestock (million)
1977	2
1980	6
1989	10
2000	18

Source: "Grapes of Wrath in Inner Mongolia," report from the U.S. Embassy in Beijing, May 2001.

reducing the amount of rainfall recycled into the interior of the continent. The Yangtze River basin, for example, which occupies much of southern China, has lost 85 percent of its original tree cover. In these circumstances, when moisture-laden air masses move inland from the sea, the rainfall they produce quickly runs off, returning to the sea. When this land was heavily forested, most of the rainfall was retained and evaporated either directly into the atmosphere or indirectly through the transpiration of the trees, to be carried further inland. As Wang Hongchang of the Chinese Academy of Social Sciences points out, the diminished capacity of the deforested land to recycle water inland may be reducing rainfall in the northwestern interior of China.[24]

China is incurring another costly ecological deficit as the use of water for irrigation, industry, and residential use climbs, exceeding aquifer recharge. When the rising demand for water approaches the sustainable yield of aquifers, governments can avoid overpumping by investing in efforts to stabilize population and by raising water productivity. Unfortunately, the sustainable-yield threshold of aquifers is usually ignored. As a result, water tables are falling throughout the northern half of China as pumping exceeds the natural recharge from precipitation. As the water levels fall, the springs that feed streams dry up. And then the rivers they feed go dry. Lakes disappear. In the northern half of China, thousands of lakes have vanished over the last few decades.[25]

A World Bank study of key river basins that make up much of the North China Plain—the Hai, which contains both Beijing and Tianjin, two of China's largest industrial cities; the Yellow, which originates on the Tibet-Qinghai plateau and eventually empties into the Yellow Sea; and the Huai, the next river basin south—found that the three together have an annual deficit of 37 billion tons of water.[26]

Assuming 1,000 tons of water to produce 1 ton of grain, this water deficit is equal to 37 million tons of grain, which at current consumption levels is enough to feed 111 million Chinese. Stated otherwise, 111 million Chinese are being fed with grain produced with the unsustainable use of water. Not only is this water deficit large, but it is growing progressively larger. With virtually all water now spoken for in northern China, the growing demand for water in cities and industry is satisfied by taking irrigation water from agriculture. For example, rice production is being phased out in the region surrounding Beijing and farmers are shifting to less water-intensive crops simply because the water is needed for the city, which now has 10 million people.[27]

The first three ecological deficits—overplowing, overgrazing, and overcutting—are destroying the vegetation that protects the soils of China. The fourth—overpumping—is drying out the land. Water shortages also make any water-dependent reclamation efforts, such as tree planting, more difficult, accelerating the desertification process.

In addition to the four ecological deficits just described, the worldwide rise in temperature may also be contributing to the desertification of China. Higher temperatures appear to be raising evaporation rates and drying out the country's interior. Warmer winters both reduce snow cover and lead to the earlier loss of snow cover in the spring, which may explain why the dust storms have started earlier in recent years. Simply stated, China may also be battling the effects of global warming.[28]

China's ecological deficits reflect three dangerous weaknesses of markets: their inability to recognize and respect the sustainable-yield thresholds of natural systems; their inability to value nature's services, such as the role of natural vegetation in protecting the land; and their inability to incorporate the indirect environmental costs of various economic activities, such as overplowing. Driven by a combination of population and income growth, these ecological deficits are setting the stage for an ecological meltdown in China on a scale that has no historical precedent.

Spreading Deserts: The Response

Until recently, coping with desertification was left largely to provincial and local governments. But the dust storms reaching Beijing in the last few years have gotten the attention of Chinese leaders. Now the federal government is beginning to commit substantial amounts of resources. The Ministry of Forestry has been designated the lead agency in the effort to arrest the spreading deserts. For example, the government is now paying farmers in the threatened provinces to abandon grain production and to plant their land in trees. In 2000 and 2001, 1.5 million hectares of cropland were planted to trees. An estimated 2 million hectares are scheduled for conversion in 2002. By 2010, 7 million additional hectares of cropland are to be covered with trees. Altogether these 10.5 million hectares represent more than a tenth of China's grainland.[29]

Halting the advancing sand dunes will not be easy. Research by Chinese scientists indicates that the millions of sheep and goats traversing the land not only strip it of vegetation but also loosen the soil through their constant trampling, leaving it particularly vulnerable to wind erosion. Without the sheep and goats, rainfall interacting with the soil forms a protective crust that helps prevent the blowing of the soil.[30]

Efforts to arrest the desertification and to reclaim the land for productive uses involve planting the land to desert shrubs that help stabilize the dunes and, in many situations, banning sheep and goats entirely. In Helin County, south of the Inner Mongolian capital of Hohhot, such a strategy is beginning to yield results. The planting of desert shrubs on cropland, which was abandoned earlier because sand dunes were forming, has now stabilized the county's first 7,000-hectare reclamation plot. The second and third 7,000-hectare reclamation efforts are under way, with a fourth to be launched before the end of 2002.[31]

A plan to deal with desertification in the threatened parts of China is complicated by the prevalence of poverty in these same regions. The situation calls for a carefully formulated strategy that will lead both to environmental stability and to economic improvement.

The strategy for Helin County, with a population of 150,000 people, is to shift the emphasis from sheep and goats to dairy cattle, increasing from 30,000 dairy animals to 150,000 over the next five years, while gradually reducing the number of sheep and goats. In contrast to the sheep and goats, which range across the landscape consuming any available vegetation, the cattle will be stall-fed, eating cornstalks, straw from the spring wheat crop, and the harvest from a drought-tolerant leguminous forage crop resembling alfalfa, which is growing on reclaimed land. Local officials estimate that this strategy will double incomes within the county during this decade.[32]

The successes in arresting and reversing the spread of the desert tend to be local and small-scale, typically in a village, a cluster of villages, or an oasis. Wang Tao, Deputy Director of CAREERI, describes two such cases. First, after rehabilitation of the Naiman Banner experimental plot in Inner Mongolia, the village's 1,000 hectares of shifting sand land decreased to 330 hectares, vegetation cover increased from 10 percent to 70 percent, the grain harvest climbed from 150 tons to 450 tons, and per capita income increased from 174 yuan per year to 1,290 yuan.[33]

Second, in a project in Ningxia Province, rehabilitation brought the 4,822 hectares of desertifying land under control. Of this, 667 hectares of shifting sand land was transformed into woodland. Vegetation cover overall increased from 30 percent to 50 percent. The grain harvest increased from 139 tons to 219 tons. Per capita income increased from less than 500 yuan per year to 1,175 yuan.[34]

Some remediation and reclamation efforts are working. Others are not. One of the difficulties with farmers planting trees to stabilize the remaining soil is that often there is not enough soil left to support the trees. The result is mortality rates that sometimes reach 80 percent in the first year. Another disadvantage is that the dust storms are concentrated in the early months of the year, from January into early May, when the deciduous trees planted as windbreaks lack the foliage needed to slow the wind.[35]

One weakness of having the Ministry of Forestry manage the land reclamation effort is that it focuses on planting trees. While tree planting has a key role to play, there are doubts as to whether it should be the core strategy. Yet in mid-May 2002, the government announced that it would be investing $12 billion in a decade-long tree planting effort to reduce wind erosion and the spread of deserts. This ambitious planting program includes all regions of the country.[36]

All too often, efforts to arrest desertification focus on the symptoms rather than the causes. There is in Beijing something of a "great wall" mentality, one that emphasizes planting a belt of trees to protect Beijing and nearby Tianjin—two of China's largest cities—from dust storms. Shi Yuanchun, a soil scientist at the China Academy of Sciences challenges this approach. "Putting hundreds of millions of dollars into the Beijing-Tianjin Sand Prevention and Forest Belt Project and ignoring the major sand source regions is...practicing self deception," he wrote. This planting of trees around Beijing is being justified partly in terms of wanting to green the city and clean the air before the city hosts the Olympics in 2008.[37]

A similar situation exists in Lanzhou, where the mountainsides that line the valley en route to the airport—mountains that have never been forested—are being covered with newly planted trees. To enhance their chances of survival, the seedlings are irrigated with large overhead sprinklers using powerful pumps to draw the water from the Yellow River far below. There are widespread doubts as to whether this prodigal use of scarce water resources to make the drive to the airport more scenic warrants the huge drain on fiscal resources it represents.

Planting trees anywhere in China is an obvious environmental plus if the trees survive. They hold the soil and retain rainfall, reducing runoff and flooding. But unless the root causes of desertification—particularly the overgrazing and overplowing in the west and north—are addressed directly, then the tree belts bordering Beijing will not protect it from dust storms.

The prevailing scientific opinion appears to be that the key to arresting the spread of deserts is to relieve the pressure posed by China's 290 million sheep and goats. Owners are being encouraged to reduce their flocks by 40 percent. In parts of the country where wealth is measured not in annual income but in the number of livestock owned and where a majority of families are living under the poverty line, such cuts are not easy. Flocks are indeed being reduced in many areas, but the reductions appear to be more like 20 percent than the suggested 40 percent. Whether even a 40-percent reduction in herd size is sufficient to arrest the desertification of land is doubtful.[38]

Arresting desertification may depend more on grass than trees—in terms of both permitting existing grasses to recover and planting grass in areas that have been denuded. The problem, as one observer has noted, is that there is a Ministry of Forestry but no Ministry of Grass. One of the common components of successful land reclamation efforts involves not merely reducing the number of sheep and goats that traverse the land, but banning them entirely until the indigenous grasses and shrubs can recover. The plan to plant marginal cropland in trees involves paying farmers 1,500 kilograms of grain per year for five years for each hectare they convert from grain to trees, providing the tree survival rate is 80 percent at the end of the first year. They will also receive a small cash allowance. This helps correct some of the mistakes of overplowing, but it does not deal with the overgrazing issue.[39]

Qu Geping, the Chairman of the Environment and Resources Committee of the National People's Congress, has said that the remediation of land in the areas where it is technically feasible would cost $28.3 billion. This dwarfs anything the government has allocated to date, raising questions about whether the government, focused on making the capital city "green" for the 2008

Olympics, has fully recognized yet the scale of the effort needed to win the war with the advancing deserts.[40]

The National Costs of Failure

The fallout from the dust storms is social as well as economic. Millions of rural Chinese may be uprooted and forced to migrate eastward as the deserts claim their land. Wang Tao reports that desertification is already producing refugees in Gansu, Inner Mongolia, and Ningxia Provinces. A preliminary Asian Development Bank assessment of desertification in Gansu Province reports that 4,000 villages risk being overrun by drifting sands.[41]

The U.S. Dust Bowl of the 1930s forced some 3 million "Okies" and other refugees to leave the land, many of them heading west from Oklahoma, Texas, and Kansas to California. But the dust bowl forming in China is much larger than that in the United States, and during the 1930s the U.S. population was only 150 million—compared with China's 1.3 billion. Whereas the U.S. migration was measured in the millions, China's may measure in the tens of millions. And as a U.S. embassy report noted, "unfortunately, China's 21st-century 'Okies' have no California to escape to—at least not in China."[42]

Not only are spreading deserts disrupting air travel, as noted earlier, but sand dunes are also encroaching on highways and railways. Along the railroad from Hohhot, the capital of Inner Mongolia, to Lanzhou in Gansu Province, stones can be seen piled in fences two or three feet high to serve as sand traps. These are designed to prevent the drifting sand from covering the railroad, much as highway departments in the United States use snow fences along highways to prevent drifting snow from disrupting road transportation.[43]

The ecological deficits building in China suggest that not only will this nation continue to lose land to invading deserts, but that the loss will be greater each year. These expanding deserts affect every facet of life in China, including food production, transportation, and population distribution. As noted earlier, the government is planning to convert 10.5 million hectares of cropland to trees during this decade, which is roughly one tenth of China's cropland.

Two other trends are also shrinking the cropland area. In addition to the planned conversion of cropland to trees, other low-productivity land is simply being abandoned. A combination of low productivity and a reduction in government grain support prices has eliminated any profit on the more marginal land, compelling many farmers to look for jobs off the land. Cropland is also being abandoned because it is covered by drifting sand or overrun by sand dunes. For example, an Asian Development Bank document reports that in one area of Gansu, 133,000 hectares of cropland have been abandoned to drifting sand. These three trends are combining to shrink China's cropland base.[44]

As rangeland turns to desert, the number of livestock that can be supported will diminish. In areas where cattle are being favored over sheep and goats, flocks of the latter will be substantially reduced. All in all, China's pastoral economy and its animal population will likely shrink dramatically in the years ahead either because livestock numbers are reduced by policy as efforts to control desertification acquire momentum or because rangelands are simply overrun by deserts.

In the end, the desertification of China is diminishing the country's food supply. As marginal cropland is systematically converted to trees or is abandoned for economic reasons, and as cropland and rangeland are abandoned to advancing deserts, the country's agricultural land area is shrinking.

The loss of productive land to desertification, along with the depletion of aquifers and the diversion of irrigation water to cities and industry, makes it increasingly difficult to expand food production. These trends are combining with economic developments—including the lowering of grain support prices in recent years, the rising wages in off-farm employment that pull labor from agriculture, and the shift to more intensive cropping, such as vegetable production, to reduce China's grain harvest.

After increasing more than fourfold from 90 million tons in 1950 to 392 million tons in 1998, China's grain production has dropped, falling to 338 million tons in 2001. (See Figure 1–1.) Even as China loses cropland, its grain consumption is rising by roughly 4 million tons each year as population expands and as people continue to use more grain-dependent livestock, poultry, and fish products. With some improvement in rainfall, the grain harvest

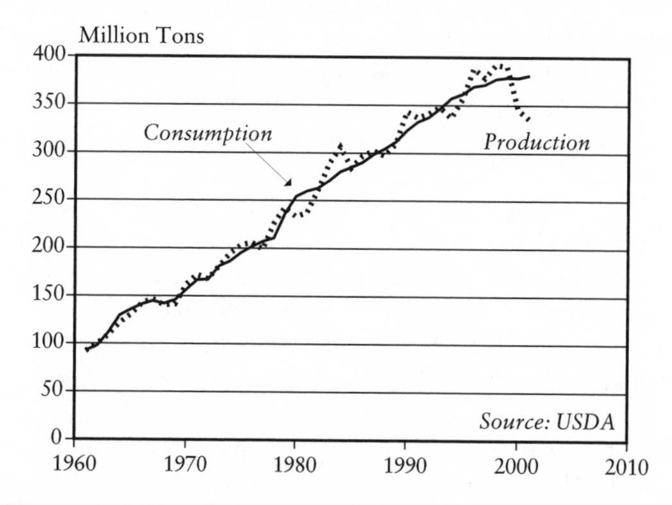

Figure 1–1. *Total Grain Production and Consumption in China, 1961–2001*

could recover to 350 million tons in 2002. Even so, with consumption now approaching 390 million tons, this will make China's third consecutive year with a shortfall of around 40 million tons. Thus far, this deficit has been filled by drawing down stocks. But if the deficit continues, China will be forced at some point in the not-too-distant future to turn to the world grain market.[45]

The Worldwide Effect of Failure

If dust storms continue on the scale and with the frequency of the last few years, they will continue to affect nearby countries, including North and South Korea, Japan, and eastern Russia. As noted earlier, they are no longer merely a nuisance; they are now taking an economic toll, especially when they disrupt transportation and close schools and factories.

Another potential consequence of desertification's shrinking the inhabitable area while the population continues to expand is that migration could change from internal to international. At some point, if more and more Chinese are squeezed into an ever-smaller area, the pressure to migrate abroad will intensify. Exactly where the dust bowl refugees would migrate to remains to be seen.

But perhaps the most immediate consequence of a failure by China to reverse the desertification of its landscape will be its effect on world grain markets, and thus on world food prices. In 1972, the Soviets decided after a poor harvest that rather than slaughter some of their herds, as they had done in similar situations in the past, they would simply import grain to offset the shortfall. As Soviet wheat imports abruptly climbed from 3 million tons in 1971 to 15 million tons in 1972, the world wheat price per bushel leapt from $1.90 in 1972 to $4.89 in 1974.[46]

If China were to import even 10 percent of its total grain supply, or 40 million tons per year, it would be good news for farmers in exporting countries because world grain prices would likely climb off the top of the chart. The bad news is that if grain prices doubled, they could destabilize governments in low-income countries that rely heavily on imports, such as Algeria, Egypt, Indonesia, Iran, or Mexico. This could also impoverish more people in a shorter period of time than any event in history. It would create a world food economy dominated by scarcity rather than by surpluses, as has been the case over most of the last half-century.[47]

And what if China were to someday consider importing 20 percent of its grain, which is still far less than the 40 percent or more imported by Algeria, Egypt, or Iran, or the 70 percent imported by Japan, for example? This would mean importing 80 million tons of grain. But where would such a large quantity of grain come from?

Many of the countries that already import a large share of their grain are still raising their imports. If China moves into the world grain market in a major way, as is now a distinct possibility, then importing countries will be competing for inadequate supplies of exportable grain. In such a world, China would probably fare better than most simply because it has an export surplus with the United States in excess of $80 billion a year. At a price of $125 per ton of grain, $1 billion will buy 8 million tons of grain. China thus could easily afford 80 million tons of grain using only $10 billion of its trade surplus with the United States. But how would the low-income countries that lack such purchasing power fare in world markets?[48]

No country has ever faced a potential ecological catastrophe on the scale of the dust bowl now developing in China. Merely grasping its dimensions and consequences poses a serious analytical challenge. Fashioning an effective response is even more demanding. At this point, there is no funded plan in place or on the drawing board that will halt the advancing deserts.

China is taking some of the right steps, such as paying farmers to plant trees on fragile soils. But it still has a long way to go in order to reduce livestock numbers to a sustainable level and to stabilize aquifers. If the government is serious about reversing desertification, it will have to commit a massive amount of human and financial resources. In terms of national priorities, it will have to decide whether to use public resources to complete the Three Gorges Dam and build the costly proposed south-north water diversion project or, instead, to halt the deserts that are marching southward and eastward and that could eventually occupy Beijing. Whether China can effectively respond to this threat may offer some insight as to whether the world as a whole will be able to arrest the deteriorating relationship between the global economy and the earth's ecosystem before it leads to economic decline.

Assessing the Food Prospect

Lester R. Brown

Throughout most of human existence, the scale of economic activity was small relative to the size of the earth's ecosystem. But over the last century this has changed. In 1900, global economic output totaled $2.4 trillion. In 2001, it was $46 trillion, an expansion of 19-fold. The world economy is now so large that its growth in the year 2000, a single year, exceeded that of the entire nineteenth century.[1]

The growth in population and in individual incomes, the two elements of this phenomenal growth, have both escalated over the last half-century. Population went from 2.5 billion at mid-century to 6.1 billion in 2001. Those of us born before 1950 are members of the first generation to witness a doubling of world population during our lifetimes. Stated otherwise, the growth in world population since 1950 is greater than that during the preceding 4 million years since our early ancestors first stood upright.[2]

Individual income climbed from $2,582 in 1950 to $7,454 in 2001, nearly tripling. Despite the extraordinary growth in the global economy over the last half-century, 1.2 billion people, one fifth of humanity, still live in abject poverty. The average income in the 20 richest countries is 37 times that of the poorest 20 countries.[3]

Since 1950, the growth in individual incomes has accounted for slightly over half of the economic expansion. Between 1950 and 2001, population grew by 146 percent and individual incomes by 188 percent. Population growth has come to a halt in 32 countries. In these nations, births and deaths are essentially in balance. Scores more want to stabilize their populations. No country, however, has stabilized individual consumption, however high it may already be.[4]

Although the global economy expanded nearly sevenfold from 1950 to 2001, the earth's ecosystem did not expand. The amount of water produced by the hydrological cycle is essentially the same today as it was in 1950. The capacity of oceanic fisheries to supply fish has not increased. Nor has the capacity of rangelands to support livestock or that of forests to supply wood for fuel, lumber, and paper. The earth's capacity to fix carbon is not increasing and may have decreased. Its capacity to absorb waste has not changed.

While the capacities of the earth's natural systems have not increased—and in many cases have diminished—the demands being placed on them have risen dramatically. World water use has tripled since 1950. The oceanic fish catch has expanded nearly fivefold. The pressures on forests to supply fuel, lumber, and paper have multiplied severalfold. Paper use has increased sixfold. Pressures on rangelands have intensified as the demand for beef and mutton has nearly tripled since 1950.[5]

In much of the world, the demands placed on natural systems have become excessive, leading to their deterioration and, in some locations, their collapse. The relationship between the global economy and the earth's ecosystems is an increasingly stressed one. Many of the stresses, including expanding deserts and increasingly frequent dust storms, rising temperature, falling water tables, eroding soils, collapsing fisheries, melting glaciers, and rising seas directly affect the food prospect.

These signs of stress, these trends of deterioration, are in large measure the result of market failures. The market has many strengths, but it also has some weaknesses that were not evident when the human enterprise was much smaller.

The market economy has brought a wealth to the world that our ancestors could not even have imagined. It allocates resources among competing uses, it balances supply and demand, and it facilitates the specialization that underpins the productivity of modern economies. But as the economy expands, the market's weaknesses are beginning to surface. Three stand out: its lack of respect for the sustainable-yield thresholds of natural systems, its inability to value nature's services properly, and its failure to incorporate the indirect costs of providing goods and services into their prices.

Soil: Surplus to Deficit

In some ways, the most fundamental ecological deficit the world faces is the loss of soil through wind and water erosion. This loss of an invaluable natural capital asset and the associated loss of land productivity are spreading as pressure on the land intensifies.

Soil erosion is not only widespread, but it is not reversible in any meaningful human time frame. Once nutrient-rich topsoil is lost, the capacity of the land to store the nutrients and the water that plants need to sustain growth is greatly diminished.

Soil scientists have assessed the risk of human-induced desertification—land that is losing its productivity as a result of human activity—and the number of people affected by it. Using four categories of risk of desertification—low, moderate, high, and very high—

they estimate that 11.9 million square kilometers are at very high risk. (See Table 1–3.) They argue that the land in the very high risk category should be the focus of policymakers because if measures are not taken to protect it soon, its productivity may be lost forever.[6]

The researchers call each of these categories a "desertification tension zone," and they are particularly concerned with the very high risk zone both because this area could turn to desert so quickly and because 1.4 billion people live there. For many of these people, their land is their livelihood.[7]

The Sahelian region of Africa, the broad band that stretches across the continent between the Sahara and the rainforest to the south, is one of the areas in serious trouble, slowly turning into desert. U.N. Secretary-General Kofi Annan reports that unless the desertification of this region is halted, within the next 20 years some 60 million people will be leaving the region—refugees from the desert.[8]

Table 1–3. *Land at Risk of Human-Induced Desertification*

Degree of Risk	Area at Risk (million square kilometers)
Low risk	7.1
Moderate risk	8.6
High risk	15.6
Very high risk	11.9
Total	49.2

Source: Hari Eswaran, Paul Reich, and Fred Beinroth, "Global Desertification Tension Zones," in D. E. Stott, R. H. Mohtar, and G. C. Steinhardt (eds.), *Sustaining the Global Farm* (2001), pp. 24–28.

Soil erosion is not new. What is new is the rate of erosion. New soil forms when the weathering of rock exceeds losses from erosion. Throughout most of the earth's geological history, the result was a gradual, long-term buildup of soil that could support vegetation. The vegetation in turn reduced erosion and facilitated the accumulation of topsoil. At some recent point in history, probably within the last century or two, this relationship was reversed—with soil losses from wind and water erosion exceeding new soil formation. The world now is running a soil deficit, one that is measured in billions of tons per year and that is reducing the earth's productivity. In China, as noted earlier, and in scores of other countries, the loss of soil is draining the land of its productivity.

In some areas, such as the flat fertile plains of Western Europe and the rice paddies of Asia, soils are stable. In others, including arid and semiarid regions, such as the Great Plains of the United States, most of Africa, Central Asia, and parts of northwestern China, land is vulnerable to wind erosion. Wherever there is sloping land, water erosion is a potential and often increasingly serious problem. In mountainous countries, such as Indonesia, Nepal, and Peru, sloping land can quickly lose its topsoil to water erosion.

In the early 1990s, some 250 scientists from 21 ecological regions concluded that 2 billion hectares of land, including cropland, rangeland, and woodland, had been degraded to some degree. This is roughly three times the 700 million hectares planted to grain worldwide. The overwhelming share of this land—84 percent—suffered from the erosion of soil, either by wind or by water.[9]

Scores of countries, mostly developing ones, are suffering a decline in inherent land productivity because of erosion. This does not necessarily mean that the harvest is declining, because in many situations advances in technology are more than offsetting the gradual loss of topsoil. But the cultivation of land that is losing its topsoil eventually becomes uneconomic, regardless of the level of technology.

As soil erodes, the land initially suffers from declining productivity, and eventually it may be abandoned. A dozen or so U.S. studies analyzing the effect of soil erosion on corn and wheat yields found that the loss of an inch of topsoil typically lowered yields by 6 percent. If the erosion continues indefinitely, it will eventually reduce the productivity of the land to the point where it can no longer be economically farmed. When the cost of producing food exceeds its value, the land is abandoned. For farmers, the cost of this ecological deficit is abandonment of their farms. For society, it represents a loss of natural capital that cannot be replaced in any meaningful time frame.[10]

Ohio State University agronomist Rattan Lal estimates that soil erosion has reduced Africa's grain harvest by 8 million tons, or roughly 8 percent. He projects this loss will double to 16 million tons by 2020 if soil erosion is not reduced. So Africa is projected to lose, in effect, the capacity to feed 80 million people at African levels of consumption during a period when its population is projected to increase by 288 million.[11]

Given the fastest population growth of any continent and some of the world's worst soil erosion, it comes as no surprise that grain production per person in Africa has been declining for the last few decades. While grain output per person in Europe, where population has stabilized and soil erosion is minimal by comparison, has nearly doubled over the last 40 years, in Africa it has fallen by nearly one fifth. (See Figure 1–2.) Of even more concern, there are no shifts in national population and agricultural policies currently in prospect in Africa to reverse this deteriorating food situation.[12]

U.S. farmers have also suffered from land mismanagement. Despite their experience with the Dust Bowl in the 1930s, a new generation of farmers was again overplowing in the 1970s in response to record high world grain prices. As a result, soil erosion increased sharply. By the early 1980s, the United States was losing over 3 billion tons of topsoil a year, an amount equal to the topsoil on 1.2 million hectares (3 million acres). This would produce 7 million tons of grain, enough, at average world consumption levels, to supply 21 million people.[13]

The erosion in the 1980s, mostly from water, was concentrated in the midwestern Corn Belt, whereas the earlier erosion of the Dust Bowl in the 1930s, mostly from wind, was concentrated in the Great Plains. In 1985, the Congress, with strong support from environmental groups, created the U.S. Conservation Reserve Program

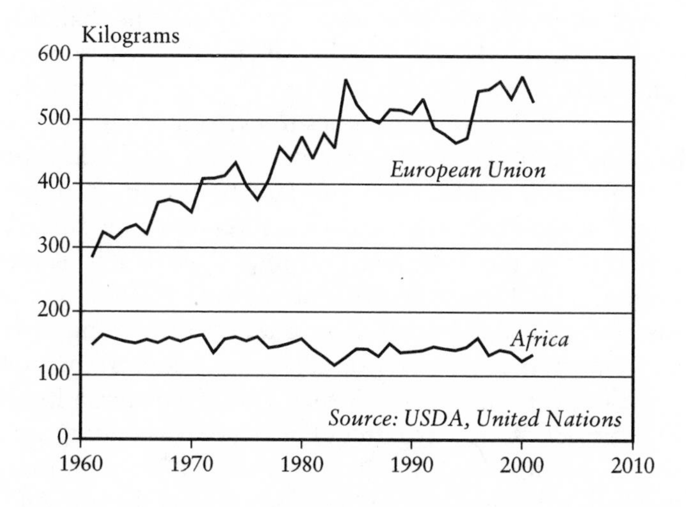

Figure 1–2. *Grain Production Per Person in Africa and the European Union, 1961–2001*

(CRP), which paid farmers to plant highly erodible cropland with grass or trees under 10-year contracts (most of which have been renewed). Within a few years, the CRP had removed some 14 million hectares (35 million acres) of cropland, nearly one tenth of the U.S. cropland total, from production. Of this land, roughly half should never have been plowed in the first place because it was so erodible. The other half could be brought back into production, if needed, with the proper soil management techniques. (Interestingly, the one tenth of cropland being converted to grass or trees is roughly the same as the share of China's cropland that is slated for conversion to trees during this decade.)[14]

Other countries that are also pulling back include Algeria, which is fighting a losing battle to protect its grainland as the desert moves northward. As a result, it has decided to convert the southernmost 20 percent of its grainland to permanent crops, either orchards or vineyards, as it tries to maintain agriculture and halt the advance of the desert. Whether or not this will succeed remains to be seen.[15]

There are few opportunities to expand production to new cropland to offset these losses. As the world demand for food has tripled over the last half-century, it has forced agriculture into areas that should not be plowed. Perhaps the most dramatic example of this is Kazakhstan. This former Soviet republic was the site of the vast Virgin Lands project during the 1950s, an initiative its supporters promised would expand grain production sufficiently to make the Soviet Union an agricultural superpower. Within a matter of years, the expanding area of grassland plowed and planted to wheat in Kazakhstan surpassed the wheat-growing area of Canada and Australia combined. It was a massive effort, but one that was destined to fail. From roughly 1960 to 1980, Kazakh farmers cultivated some 26 million hectares of grain. But by 1980 wind erosion was reducing yields to where farmers were abandoning their land because it was no longer economic to farm. By 2000, the area in grain had fallen to less than 13 million hectares. Within two decades, Kazakhstan had abandoned half of its grainland, an area equal to Canada's wheatland. Wheat yields on the remaining land average scarcely 1 ton per hectare, only a fraction of the 7 tons per hectare of France, Western Europe's largest wheat producer.[16]

Despite the history of overplowing experience in key countries, there are still a few high-risk expansion efforts under way. One consists of replacing tropical rainforests in Indonesia and Malaysia with palm oil plantations. Although this is producing cheap palm oil, it is devastating the biological diversity of the region, and without any assurance that these exposed tropical soils will sustain cultivation over the long term.[17]

A far more ambitious effort is under way in Brazil as farmers plow the cerrado—a vast, semiarid savannah that is to the south and west of the Amazon basin. This land has helped Brazil become the world's second-ranking soybean producer, after the United States. The excitement within Brazil at this region's potential is remarkably similar to that displayed by the Soviets during the Virgin Lands Project in Kazakhstan some 45 years ago. Only time will tell whether the newly plowed cerrado will sustain cultivation over the long term.[18]

The Fast-Growing Water Deficit

While the soil deficit is growing slowly, the water deficit is growing rapidly. The world water deficit—historically recent, largely invisible, and growing fast—may be the most underestimated resource issue facing the world today. Because it typically takes the form of aquifer overpumping, the resulting fall in water tables is not visible. Unlike shrinking forests or invading sand dunes, falling water tables cannot be readily photographed. They are often discovered only when wells go dry.

In round numbers, 70 percent of all the water pumped from underground or diverted from rivers worldwide is used for irrigation, 20 percent is used by industry, and 10 percent goes to residences. But the demand for water in industry and for residential purposes is growing even faster than population, putting a squeeze on the amount available for agriculture.[19]

In some 18 countries, population growth has reduced the fresh water supply per person to less than 1,000 cubic meters per year, the minimal amount needed to satisfy basic needs for drinking, hygiene, and food production. By 2050, U.N. population projections show that 39 countries, with 1.7 billion people, will be experiencing such water deprivation.[20]

For most ecological deficits, we do not have a global estimate of their size. But for water we do. In her book *Pillar of Sand*, Sandra Postel, using data for India, China, the United States, North Africa, and Saudi Arabia, estimated the annual water deficit in terms of aquifer overpumping at over 160 billion tons per year. Using the rule of thumb of 1,000 tons of water to produce 1 ton of grain, this would be enough to produce 160 million tons of grain. With current world grain consumption of 300 kilograms per person, this would feed 533 million people. Stated otherwise, 533 million of us, out of the world population of 6.1 billion, are being fed with grain that is produced with the unsustainable use of water.[21]

The world water deficit is concentrated in China, the Indian subcontinent, the Middle East, North Africa, and North America. This problem is historically recent, a product of the tripling of world water usage since 1950 and the spreading use of powerful diesel and electrically driven pumps. When the pumping of water from wells depended on human or animal power, the amount pumped was limited, but now with powerful mechanically driven pumps, aquifers can be depleted in a matter of years.[22]

In a world where the demand for water continues its steady growth while the sustainable yield of aquifers is essentially fixed, the deficits grow larger year by year. The longer that governments delay in addressing this issue, the larger the annual deficit becomes, the faster water tables fall, and the more difficult it is to deal with. Scores of countries are now experiencing water deficits—from smaller ones like Iran or Yemen to the world's most populous country, China. (See Table 1–4.)[23]

Iran, a country of 70 million people, is facing an acute shortage of water. Under the agriculturally rich Chenaran Plain in northeastern Iran, the water table was recently falling by 2.8 meters a year. But the cumulative effect of a three-year drought and the new wells being drilled to supply the nearby city of Mashad, one of Iran's largest, dropped the aquifer by an extraordinary 8 meters in 2001. Villages in eastern Iran are being abandoned as aquifers are depleted and wells go dry, generating a flow of water refugees.[24]

In Yemen, which has a population of 17 million, World Bank data indicate that the water table under most of the country is falling by roughly 2 meters a year as water use far exceeds the sustainable yield of aquifers. World Bank official Christopher Ward observes that "groundwater is being mined at such a rate that parts of the rural economy could disappear within a generation." In the basin where the capital, Sana'a, is located, the water table is reportedly falling at 6 meters (nearly 20 feet) per year. The Bank estimates that the aquifer will be

Table 1–4. *Selected Examples of Aquifer Depletion*

Country	Region	Description of Depletion
China	North China Plain	Water table falling by 2–3 meters per year under much of the Plain. As pumping costs rise, farmers are abandoning irrigation.
United States	Southern Great Plains	Irrigation is heavily dependent on water from Ogallala aquifer, largely a fossil aquifer. Irrigated area in Texas, Oklahoma, and Kansas is shrinking as aquifer is depleted.
Pakistan	Punjab	Water table is falling under the Punjab and in the provinces of Baluchistan and North West Frontier.
India	Punjab, Haryana, Rajasthan, Andhra Pradesh, Maharashtra, Tamil Nadu, and other states	Water tables falling by 1–3 meters per year in some parts. In some states extraction is double the recharge. In the Punjab, India's breadbasket, water table falling by nearly 1 meter per year.

depleted by the end of this decade. In the search for water, the Yemeni government has drilled test wells in the basin that are 2 kilometers (1.3 miles) deep—depths previously associated only with the oil industry. But they have failed to find water. Those living in Yemen's capital may soon be forced to relocate within the country or to migrate abroad, adding to the swelling flow of water refugees.[25]

Table 1–4 *continued*

Iran	Chenaran Plain, northeastern Iran	Water table was falling by 2.8 meters per year but in 2001 drought and drilling of new wells to supply nearby city of Mashad dropped it by 8 meters.
Yemen	Entire country	Water table falling by 2 meters per year throughout country and 6 meters a year in Sana'a basin. Nation's capital, Sana'a, could run out of water by end of this decade.
Mexico	State of Guanajuato	In this agricultural state, the water table is falling by 1.8–3.3 meters per year.

Source: See endnote 23.

In China, water tables are falling virtually everywhere that the land is flat. Under the North China Plain, which produces at least one fourth of the country's grain, the fall in the underground water table of 1.5 meters (5 feet) per year of the early 1990s has recently increased to 2–3 meters per year in some areas. In many parts of China, well drilling, either the deepening of wells or the drilling of new ones, has become a leading industry.[26]

The big three grain producers—China, the United States, and India, which together account for nearly half of world output—depend on irrigation in varying degrees. In China, 70 percent of the grain produced comes from irrigated land. In India, the figure is 50 percent, and in the United States, 15 percent. In each country, water tables are falling in key agricultural areas. When farmers lose their irrigation water, whether from aquifer depletion or diversion to cities, and return to rainfed farming, they typically experience a drop in yields of one half or so.[27]

Historically, water shortages were local, but in an increasingly integrated world economy, these shortages can cross national boundaries via the international grain trade. Water-scarce countries often satisfy the growing needs of cities and industry by diverting water from irrigation and importing grain to offset the resulting loss of production. Since 1 ton of grain equals 1,000 tons of water, importing grain is the most efficient way to import water.[28]

Although it is often said that future wars in the Middle East are more likely to be fought over water than oil, the competition for water in the region seems more likely to take place in world grain markets. This can be seen in Iran and Egypt, both of which now import more wheat than Japan, traditionally the world's leading importer. Imports now supply 40 percent of the total consumption of grain—wheat, rice, and feedgrains—in both countries. Numerous other water-short countries in the Middle East also import much of their grain. Morocco imports half of its grain. For Algeria and Saudi Arabia, the figure is over 70 percent. Yemen imports nearly 80 percent of its grain, and Israel, more than 90 percent.[29]

China regained grain self-sufficiency during the late 1990s in part by overpumping its aquifers, running up a huge water deficit. India is also essentially self-sufficient in grain, but it has achieved this by overpumping. Neighboring Pakistan, a country of 154 million people, is overpumping its aquifers for the same reason. Overpumping is by definition a short-term phenomenon. At some point, as aquifers are depleted, it will end.[30]

Farmers who are overpumping underground water for irrigation are facing a double squeeze. Like the rest of the economy, they face cutbacks from aquifer depletion. But in addition, they face cutbacks as irrigation water is diverted to higher value uses in industry, where the value of output per ton of water used can be easily 50 times that in agriculture. Countries seeking to create jobs and raise living standards cannot afford to use scarce water for irrigation if it deprives industry of needed water.[31]

Water deficits are already spurring heavy grain imports in numerous smaller countries, but it is unclear when they will do so in larger countries, such as India or China. Two things are obvious: the water deficits are growing larger in literally scores of countries, and they are doing so more or less simultaneously. National water shortages are not isolated events.

If countries like India, Pakistan, and China are already experiencing water deficits, what happens if their populations continue to grow as projected, demanding ever more water at a time when sharp cutbacks from aquifer depletion are imminent?

The risk is that growing water deficits in populous countries could push grain import needs above supplies in the handful of countries that export grain, triggering a rise in grain prices. It is one thing for small countries to turn to the world for much of their grain supply, but when a country like China, which consumes 390 million tons of grain per year, or India, which consumes 180 million tons, does so, it has the potential to overwhelm world grain markets, as noted earlier. If world grain prices were to double, as they did between 1972 and 1974, the ranks of those who are hungry and malnourished—estimated at 815 million—could expand dramatically as the urban poor are squeezed between low incomes and rising food prices.[32]

Humanity is moving into uncharted territory in the water economy. With the demand for food climbing, and with the overpumping of aquifers now common in industrial and developing countries alike, the world is facing a convergence of aquifer depletions in scores of countries within the next several years. As this occurs, pumping will necessarily be reduced to the rate of recharge. Unfortunately, the world has no experience in responding to water deficits on the scale now unfolding.

World agriculture, already burdened with soil and water deficits, is facing a projected addition of 3 billion people and billions of low-income people who want to move up the food chain, consuming more livestock products. These soaring demands on an agricultural system that is already ecologically stressed are leading to some basic structural changes in the world food economy.[33]

The Changing Food Economy

The biggest prospective structural changes are in the animal protein sector. During the second half of the last century, a time when population was more than doubling and incomes were nearly tripling, the world demand for animal protein was climbing. For much of this period, growth in animal protein was satisfied by turning to rangelands and oceanic fisheries. Between 1950 and 1990, beef and mutton production more than doubled and the oceanic fish catch expanded fivefold. Since then, growth in the output of these two natural systems has slowed or leveled off as demands have pressed against their sustainable-yield limits. Indeed, in many cases demand has far exceeded sustainable yields, leading to the desertification of rangelands and collapsing fisheries. Today overgrazing and overfishing are the rule, not the exception.[34]

Even as these natural systems were approaching their limits, the demand for animal protein was growing at a record rate. As it did so, the world turned to grain-fed beef, pork, poultry, eggs, milk, and farmed fish. Population growth and the strong desire to move up the food chain has driven the demand for meat steadily higher.

Indeed, except for the recession year of 1960, the world demand for meat has climbed to a new high every year since 1950. (See Figure 1–3.) Meat consumption per person more than doubled, climbing from 17 kilograms in 1950 to 39 kilograms in 2001.[35]

Wherever incomes rise, people try to diversify their diets, reducing their overwhelming dependence on a starchy food staple, such as rice, and augmenting it with meat, eggs, and milk products. This desire to move up the food chain as incomes rise appears to be innate, perhaps a genetic legacy of 4 million years as hunter-gatherers.

Given the rising demand for animal protein in diets, now principally in developing countries, the question is how to satisfy that demand most efficiently. At the first level, the advantage goes to ruminants that can convert roughage into edible forms of animal protein. The roughage may come from rangelands or from crop residues. Once the use of roughage is fully exploited,

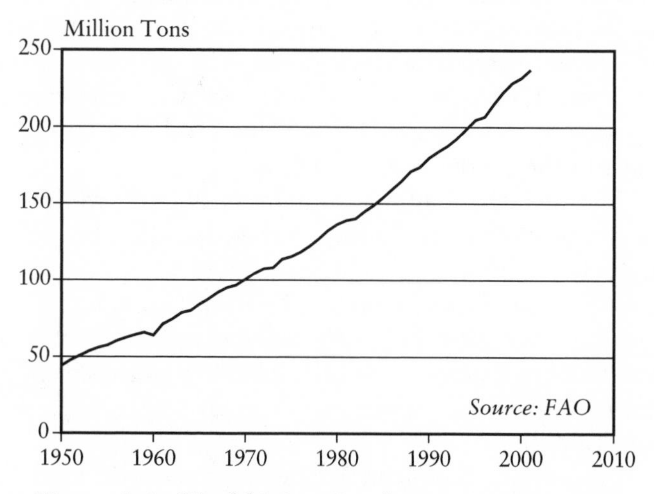

Figure 1–3. *World Meat Production, 1950–2001*

then the advantage goes to those grain-dependent forms of animal protein that are most efficient. This shift to more grain-efficient, lower-cost, animal protein sources is already under way.

There are some encouraging success stories in efficiently satisfying the hunger for animal protein. In India, for example, milk production has soared over the last few decades, spurred by local dairy cooperatives that provide a marketing link between villagers, who typically have only two to three cows, and consumers in other villages and nearby cities. Milk production in India, which has overtaken that of the United States, the longstanding leader, is based almost entirely on the use of local forage and crop residues. Little grain is fed to cattle in India.[36]

China is using a similar approach to expand beef production. In areas that produce grain, particularly those that double-crop grains, such as winter wheat and corn in east-central China, there are large amounts of crop residues—either straw from wheat or rice or the stalks from corn—that can be fed to cattle. Cattle, being ruminants, can easily convert crop residues into protein, leaving the manure to fertilize fields. The amount of beef now produced in this manner in the east-central provinces greatly exceeds that being produced on rangelands in the overgrazed northwest.[37]

The pattern of animal protein production worldwide has shifted substantially over the last decade. The growth in beef and mutton production, most of which comes from rangelands, was less than 1 percent a year from 1990 to 2001. Pork grew by nearly 3 percent, poultry production by over 4 percent. But the most rapid growth of all was in aquaculture, which expanded by 10 percent a year. (See Table 1–5.)

The variation in growth rates is explained largely by the efficiency with which various animals convert grain into protein. With cattle in feedlots, it takes roughly 7 kilograms of grain to produce a 1-kilogram gain in live weight. Growth in the number of feedlots is now minimal. For pork, the figure is close to 4 kilograms per kilogram of weight gain, for poultry it is just over 2, and for herbivorous species of farmed fish, such as carp, tilapia, and catfish, it is less than 2. The market is shifting production to the animals that convert grain most efficiently, thus lightening the pressure on soil and water resources. Health concerns are also helping to shift consumption from beef and pork to poultry and fish.[38]

Egg production is growing fast, again because laying hens can convert grain into protein rather efficiently.

Table 1–5. *Annual Growth in World Animal Protein Production, by Source, 1990–2001*

Source	1990	2001	Annual Growth
	(million tons)		(percent)
Beef	53	57	1
Pork	70	92	3
Mutton	10	11	1
Poultry	41	69	4
Eggs	38	56	4
Oceanic Fish Catch[1]	86	95	1
Aquacultural Output[1]	13	36	10

[1]Latest figures available for oceanic fish catch and aquacultural production are for 2000.

Source: Based on FAO, *1948–1985 World Crop and Livestock Statistics* (Rome: 1987); FAO, *FAOSTATS Statistics Database*, updated 28 May 2002; FAO, *Yearbook of Fishery Statistics: Capture Production* and *Aquaculture Production* (various years).

In addition, eggs are a popular source of animal protein in developing-country villages where there is no refrigeration.

Once the potential for relying on ruminants to convert roughage into food products that are edible by humans is fully exploited, then the question is how to satisfy the additional need for high-quality protein. One way of doing this is to convert grain into animal protein at varying degrees of efficiency. Another way is to supplement grain with various beans, peas, and other leguminous crops that contain high-quality protein. This can be seen, for example, in the corn-and-beans diet of Mexico or the wheat-and-lentils combination of northern India. The basic choice is whether to use the land to produce leguminous crops for direct consumption or to produce grain and convert it into animal protein.

Contrary to popular opinion, the latter may sometimes represent a more efficient use of land simply because the yield per hectare of soybeans, lentils, chickpeas, and other leguminous crops is so low compared with grain. In the United States, for example, which produces roughly 40 percent of the world corn and soybean harvests, the ratio of corn yields, at 8.7 tons per hectare, to soybean yields, at 2.7 tons per hectare, is 3.2 to 1. (The United States offers an ideal comparison between corn and soybean yields because they are grown on the same land, often in a two-year rotation.)[39]

If land is used to produce corn that is fed to a herbivorous species of fish, such as carp in China or catfish in the United States, which convert 1.5–2 kilograms of grain into 1 kilogram of live weight, or if it is fed to chickens, which use about 2 kilograms of grain to produce a kilogram of live weight, it may yield more high-quality protein than land planted to soybeans for direct consumption, for example, as tofu. In summary, the more grain-efficient forms of animal protein may not require any more land or water resources per unit of protein than legumes. At this point, whether someone consumes tofu, lentils, carp, catfish, or chicken may be less a question of the efficiency of land and water-resource use and more a question of taste.[40]

If the alternative is producing beef in feedlots, then the 7-to-1 conversion of grain to live weight of cattle is much less efficient than using land to produce beans for direct consumption. If the option is pork production and the pork is produced with table waste, as it often is in villages in China, the advantage goes to pork. But if pork is produced with grain, as is the case elsewhere and, increasingly, in China, then consuming beans directly would be more efficient.[41]

Perhaps the most impressive growth of any animal protein-producing enterprise has been fish farming in China, where a carp-polyculture has been highly successful. Over the last two decades China's aquacultural output, consisting largely of carp, has expanded from 3 million tons per year to 25 million tons. Indeed, fish-farm output in China is now double the U.S. beef output of 12 million tons.[42]

As the growth in animal protein production has shifted over the last decade or so from largely oceanic fisheries and rangelands to primarily pork, poultry, and fish farming, the pressure on croplands has intensified. Expanding protein production by feeding animals, whether fish, poultry, or pigs, means expanding grain use. At the same time, land is required by these enterprises themselves. For example, in China 5 million hectares are now devoted to fish ponds, an area equal to 6 percent of China's harvested grainland. In the United States, catfish ponds, the dominant source of U.S. farmed fish, occupy nearly 50,000 hectares (190 square miles) of land, most of it concentrated in Mississippi.[43]

As animal protein production shifts to more grain-efficient sources, it is automatically shifting to the more water-efficient sources, helping to lower the water deficit. This interaction between the expanding demand for animal protein and the need for greater efficiency in the use of grain and water is reshaping the food economy.

The Soybean Factor

Closely related to these structural changes in the world food economy is the expanding role of the soybean, perhaps the best single indicator of the world growth in animal protein consumption over the past half-century. One of the keys to the efficient conversion of grain into animal protein is the incorporation of small amounts of high-quality protein, such as that found in soybeans, into the feed ration. A modest amount of protein supplementation can boost sharply the efficiency with which grain is converted into animal protein, sometimes nearly doubling it.

Protein supplements typically come from oilseed meals, the product of crushing soybeans, cottonseed, peanuts, or sunflower seeds to extract oil. Over the last 50 years, the soybean has emerged as the world's dominant source of protein for supplementing livestock, poultry, and fish rations, exceeding all other high-protein meals combined. Between 1950 and 2001, the world soybean harvest expanded from 17 million tons to 184 million tons, a spectacular gain of nearly 11-fold. (See Figure 1–4.)[44]

The soybean, domesticated by farmers in central China some 5,000 years ago, was introduced into the United States in 1804 when Thomas Jefferson was President. After languishing as a novelty crop for a century and a half, its production began expanding rapidly following World War II, climbing from less than 6 million tons in 1950 to 79 million tons in 2001, or 43 percent of the world harvest. Brazil, now in second place, produces 24 percent of the harvest, and Argentina 16 percent. China, which once dominated the world harvest, now accounts for only 8 percent.[45]

In the United States, the harvested area of soybeans first overtook that of wheat in 1973 and that of corn in 1999. Most U.S. soybeans are produced in the Corn Belt, often in an alternate-year rotation with corn, which has a ravenous appetite for nitrogen. Since the soybean fixes nitrogen, its yields are not very responsive to the application of fertilizer. As a result, farmers get more soybeans largely by planting more land in soybeans.[46]

China, whose soybean meal use is doubling every five years as meat and egg consumption climbs, and whose direct consumption of soybeans for food is also climbing, is now the world's dominant soybean importer. Its principal supplier is the United States. The soybean connec-

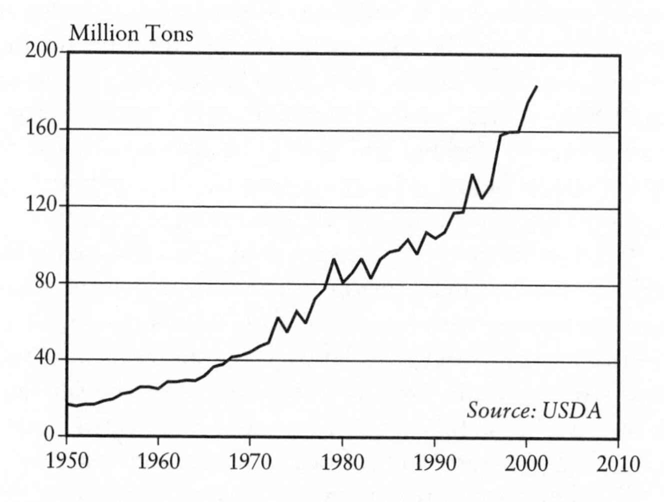

Figure 1–4. *World Soybean Production, 1950–2001*

tion between the country that gave the world the soybean and the one that made it into a world-class crop is likely to grow even stronger in the years ahead as China's appetite for animal protein continues to climb. At issue is whether soybean production can continue to expand, supporting the growth in demand for animal protein without clearing additional areas of the Brazilian cerrado, where the expansion is concentrated.[47]

Future Food Security

Future food security depends on expanding many ongoing activities such as agricultural research, agricultural extension, farm credit (especially microcredit programs designed for small farmers in developing countries), and the support prices that stabilize prices and encourage farmers to invest in land improvement. The keys to future food security now are to eliminate the soil deficit and the water deficit and to stabilize population and climate.

Reducing soil losses below the rate of new soil formation is possible, but it will take an enormous worldwide effort. Based on the experience of key food-producing countries such as China, the United States, and numerous smaller countries, easily 5 percent of the world's cropland is highly erodible and should be converted back to grass or trees before it becomes wasteland. The first step is to pull back from the fast-deteriorating margin.[48]

Wind erosion is concentrated in arid and semiarid regions, while water erosion is concentrated on sloping lands in regions with higher rainfall. Wind erosion, common on cropland and rangelands, is the source of dust and sand storms. Water erosion is the source of the silt that raises riverbeds, fills irrigation and hydroelectric reservoirs, clogs harbors, and suffocates marine ecosystems.

The key to controlling wind erosion is to keep the land covered with vegetation as much as possible and to slow wind speeds at ground level. Ground-level wind speeds can be slowed with shrubs, trees, and crop residues left on the surface of the soil. For areas that are particularly rich in wind and in need of electricity, such as northwestern China, wind turbines can simultaneously slow wind speed and provide cheap electricity. This approach converts a liability—strong winds—into an asset.

One of the time-tested methods of dealing with water erosion is terracing, as is so common with rice paddies in the mountainous regions of Asia. On land that is less steeply sloping, contour farming as in the midwestern United States has worked well.

Another tool, a relatively new one, in the soil conservation toolkit is conservation tillage, which includes both no tillage and minimum tillage. After being taught that seedbeds required plowing and careful preparation prior to planting, farmers are now learning that less tillage may be better. Instead of plowing land, then discing or harrowing it to prepare the seedbed, then planting with a seeder and cultivating row crops with a cultivator two or three times to control weeds, farmers simply drill seeds directly into the land without any preparation at all. Weeds are controlled with herbicides. This means the only tillage is often a one-time disturbance in a narrow band of soil where the seeds are inserted, leaving the remainder of the soil undisturbed.[49]

This practice, now widely used in the production of corn and soybeans in the United States, has spread rapidly in the western hemisphere over the last few decades. (See Table 1–6.) Data for crop year 1998/99 show the United States with 19.3 million hectares of land under no-till. Brazil had 11.2 million hectares, and Argentina 7.3 million hectares. Canada, at 4 million hectares, rounds out the "big four."

In the United States, the combination of retiring the highly erodible land under the CRP and shifting part of the remaining land in row crops to conservation tillage has sharply reduced soil erosion. By 2000, U.S. farmers were no-tilling 21 million hectares (51 million acres) of crops. An additional 23 million hectares were minimum-tilled, for a total of 44 million hectares of conservation tillage. This was used on 12 million hectares (30 million acres) of corn—or 37 percent of the crop. For soybeans, it was 17 million hectares—57 percent of the crop. For wheat and other small grain crops, the conservation tillage area totaled 11 million hectares (30 percent of the planted area).[50]

Once farmers begin to practice no-till, its use can

Table 1–6. *Cropland Area Under No-Till in Key Countries, 1998/99*

Country	Area
	(million hectares)
United States	19.3
Brazil	11.2
Argentina	7.3
Canada	4.1
Australia	1.0
Paraguay	0.8
Mexico	0.5
Bolivia	0.2
Others	1.1
Total	45.5

Source: Rolf Derpsch, "Frontiers in Conservation Tillage and Advances in Conservation Practice," in D. E. Stott, R. H. Mohtar, and G. C. Steinhardt (eds.), *Sustaining the Global Farm* (2001), pp. 248–54.

spread rapidly. In the United States, the no-till area went from 7 million hectares in 1990 to nearly 21 million hectares in 2000, tripling within a decade.[51]

Recent U.N. Food and Agriculture Organization reports describe the growth in no-till farming over the last few years in Europe, Africa, and Asia. In addition to reducing both wind and water erosion, and particularly the latter, this practice also helps retain water and reduces the energy needed for crop cultivation.[52]

While the soil deficit has been building over the last few centuries, the water deficit is much more recent, a product of the half-century or so since diesel and electrically powered pumps have become widely available for irrigation. And, as noted earlier, it is growing fast.

The potential disruption of world grain markets by water shortages calls for a global effort to raise water productivity, an effort similar to that launched 50 years ago regarding land. When it was realized after World War II that there was not much new land to bring under the plow, a worldwide effort was undertaken to raise land productivity. It included heavy investment in agricultural research to raise crop yields, the development of agricultural extension services to disseminate the research results to farmers, and the adoption of support prices to stabilize prices of farm commodities. This effort was highly successful, boosting world land productivity from 1.1 tons of grain per hectare worldwide in 1950 to 2.7 tons per hectare in 2001.[53]

Future food security now depends on raising water productivity not only in agriculture but in all sectors of the economy—ranging from more water-efficient household appliances to more water-efficient irrigation systems. Of all the policy steps to raise water efficiency, by far the most important is establishing a price for water that will reflect its value to society. Because water policies evolved in an earlier age, when water was relatively abundant, the world today is sadly lacking in policies that reflect reality. Raising the price of water to reflect its value would affect decisions involving its use at all levels and in all sectors. To be successful, the price should go up in concert with what some countries describe as "lifeline rates," where individual residences get the amount of water needed to satisfy basic needs at an easily affordable price. But once water consumption exceeds this minimum needs level, then the cost would escalate, thus encouraging investment in water efficiency.

The underpricing of water permeates water systems throughout the world. Some governments, such as India, heavily subsidize the use of irrigation water by providing electricity for pumping water to farmers at a nominal cost.[54]

Since 70 percent of all water that is diverted from rivers or pumped from underground is used for irrigation, investment in more water-efficient irrigation practices and technologies is central to any effective strategy to raise water productivity. In simplest terms, this means shifting from less water-efficient flood or furrow irrigation to more-efficient sprinkler and drip irrigation. Drip irrigation—now used on some 2.4 million hectares of cropland worldwide—can easily reduce irrigation water use by half while boosting yields. Its drawback is that it is much more labor-intensive. But in countries with widespread unemployment, switching to drip irrigation for many crops would simultaneously raise water productivity and employment. Although drip irrigation is not economic in all situations, there are many where it is economic but not yet used.[55]

There is also the possibility of adopting irrigation practices that use water more efficiently. In some situations, for example, rice need not be permanently flooded throughout the growing season but can be flooded periodically without any loss in yield.[56]

Cropping patterns are also being altered to favor more water-efficient crops. Both Egypt and China restrict the production of rice because of its high water requirements, favoring wheat instead. Anything that raises the efficiency of grain conversion into animal protein also raises water efficiency.

For those who are living high on the food chain—that is, who are consuming excessive amounts of fat-rich livestock products—moving "down" the food chain will both improve personal health and lower grain use and, therefore, water use. Consuming less fat also reduces obesity and the associated costs of treating obesity-related illnesses.

A third step to enhance future food security is to stabilize world population growth sooner rather than later. Current U.N. projections for 2050 range from a low projection of 7.9 billion to the high of 10.9 billion. The prospects of everyone having enough food will be greatly enhanced if the world can reach only the lower number. Even with existing populations, many developing countries do not have enough water to meet basic needs. What happens if their populations double again, as some are projected to do in the next few decades? The key now is to invest in the education of young females throughout the developing world and to improve the status of women by giving them the same ownership, inheritance, and voting rights as men. This, combined with filling the family planning gap, so that couples everywhere have access to family planning services, is the key to future food security and to making sure that people everywhere will have enough food to develop their full physical and mental potential.[57]

The other key to future food security is climate stabilization. World agriculture as it exists today evolved over 11,000 years during a period when climate was remarkably stable. If temperature and rainfall levels and patterns begin to change, agriculture as it currently exists will be out of sync with the ecosystem, forcing the need for constant adjustment as the climate system itself changes. Climate change is now the wild card in the food security deck of cards.

Facing the Climate Challenge

Lester R. Brown

The earth is getting warmer. The 15 warmest years since global recordkeeping began in 1867 have all come since 1980. Hardly a week goes by without new reports of ice melting, record temperature highs, or more destructive storms.[1]

The temperature series maintained by the Goddard Institute for Space Studies of the National Aeronautics and Space Administration (NASA) shows that the 10 months preceding June 2002 (August 2001 through May 2002) were uncommonly warm, setting several records. Temperatures for September and November were the highest ever recorded for those months in the last 134 years. Those for August, December, January, March, April, and May were the second highest on record for those months. If these record or near-record temperatures continue, then 2002 will likely set a new annual record, moving above 1998, the previous high.[2]

With emissions of carbon dioxide (CO_2), the principal greenhouse gas, continuing to rise, further increases in temperature are almost inevitable. The latest report by the Intergovernmental Panel on Climate Change (IPCC) projects that global average temperature will rise by 1.4–5.8 degrees Celsius (2.5–10.4 degrees Fahrenheit) by the end of this century. This will undoubtedly alter every ecosystem on the earth and every facet of human activity.[3]

Perhaps the most pervasive evidence of warming to date is seen in ice melting. In Alaska, where wintertime temperatures now average up to 7 degrees Fahrenheit above the norm, glaciers are retreating at an accelerating rate. A similar situation exists in the Andes. And recent data on the Himalayas indicates that glaciers there too are melting at an alarming pace.[4]

One of the concerns of scientists is that climate change will not always be a linear process. For example, if the ice in the Arctic Sea continues to melt, leaving the sea ice-free during the summers, as projected for sometime within the next several decades, the heat balance of the region could change dramatically. With the Arctic Sea largely covered with ice and snow, roughly 80 percent of the incoming sunlight is bounced back into space, while 20 percent is absorbed as heat. But an ice-free Arctic Sea during the summer would mean that 20 percent of the incoming sunlight will bounce back into space and 80 percent will be absorbed as heat. While the melting of the Arctic Sea ice does not affect sea level, a dramatic warming of the Arctic could lead to rapid melting of the Greenland ice sheet.[5]

An article in *Science* reports that if the Greenland ice sheet were to melt entirely, and this could only happen over a long period of time, it would raise sea level by 23 feet. At some point, feedback loops, such as the one just described for the Arctic, could begin to reinforce existing trends. Once certain thresholds are crossed, change can come rapidly and unpredictably—leaving a bewildered and perhaps frightened world in its wake. At issue is whether our political institutions, which could not prevent these mega-scale changes, will be able to deal with them when they occur.[6]

The Rising Costs of Climate Change

The benefits of burning fossil fuels are well known, but there are also enormous costs, many of which will be levied on future generations. Among these are changes in temperature that result in more destructive storms, rising seas, and crop-withering heat waves. More destructive storms are the product of the higher water surface temperatures, particularly in the tropical and subtropical regions, where hurricanes (typhoons in the Pacific) originate. These higher temperatures mean more energy is released into the atmosphere to drive storm systems. Rising global temperatures also mean rising seas from both thermal expansion and ice melting. Crop-withering heat waves are often accompanied by drought, with the two reinforcing each other.

At the end of 2001, Munich Re, the world's largest reinsurer (a company that helps spread risk among the various insurance companies), compiled a list of all natural catastrophes on record with insured losses of $1 billion or more. (See Table 1–7.) The first such disaster came in 1983, when Hurricane Alicia racked up $1.3 billion worth of insured damages in the United States. By the end of 2001, the list of catastrophes with insured damages of $1 billion or more had reached 34. Of these, 32 were storms, floods, and other atmospherically related events. The other two were earthquakes.

Insured damage from storms is rising for four reasons. One, more property is covered by insurance today than in the past. Two, the value of the property (as measured in dollars) has increased. Three, there is more building in coastal regions, on river floodplains, and in other high-risk areas. And four, storms are both more frequent and more powerful.

In the last 15 years, Europe has experienced a greater frequency of highly destructive winter storms. From 1987

Table 1–7. *Atmospherically Related Catastrophes with Over $1 Billion in Insured Losses, through 2001*

Year	Event	Location	Insured Losses	Economic Losses
			(billion dollars)	
1983	Hurricane Alicia	United States	1.3	3.0
1987	Winter storm	Western Europe	3.1	3.7
1989	Hurricane Hugo	Caribbean, United States	4.5	9.0
1990	Winter Storm Daria	Europe	5.1	6.8
1990	Winter Storm Herta	Europe	1.3	2.0
1990	Winter Storm Vivian	Europe	2.1	3.2
1990	Winter Storm Wiebke	Europe	1.3	2.3
1991	Typhoon Mireille	Japan	5.4	10.0
1991	Oakland forest fire	United States	1.8	2.0
1992	Hurricane Andrew	United States	17.0	30.0
1992	Hurricane Iniki	Hawaii	1.6	3.0
1993	Snow storm	United States	1.8	5.0
1993	Flood	United States	1.0	16.0
1995	Hail	United States	1.1	2.0
1995	Hurricane Luis	Caribbean	1.5	2.5
1995	Hurricane Opal	United States	2.1	3.0
1996	Hurricane Fran	United States	1.6	5.2
1998	Ice storm	Canada, United States	1.2	2.5
1998	Floods	China	1.0	30.0
1998	Hail, severe storm	United States	1.4	1.8

Table 1–7 *continued*

1998	Hurricane Georges	Caribbean United States	4.0	10.0
1999	Hail storm	Australia	1.1	1.5
1999	Tornadoes	United States	1.5	2.0
1999	Hurricane Floyd	United States	2.2	4.5
1999	Typhoon Bart	Japan	3.5	5.0
1999	Winter Storm Anatol	Europe	2.4	2.9
1999	Winter Storm Lothar	Europe	5.9	11.5
1999	Winter Storm Martin	Europe	2.5	4.0
2000	Typhoon Saomai	Japan	1.0	1.5
2000	Floods	United Kingdom	1.1	1.5
2001	Hail, severe storm	United States	1.9	2.5
2001	Tropical Storm Allison	United States	3.5	6.0

Source: Munich Re, *Topics Annual Review: Natural Catastrophe 2001* (Munich, Germany: 2002), pp. 16–17.

to 2001 the continent was battered by eight storms with insured damage of $1 billion or more. The first, in 1987, led to $3.1 billion in insured losses. In 1990 there was a clustering of four storms with damages ranging from $1.3 billion to $5.1 billion. Lothar, one of three storms to hit Europe in the winter of 1999, had insured losses of $5.9 billion and total losses of $11.5 billion, making it the most costly storm on record in Europe. Until recently, such destructive storms had been largely confined to the hurricane belt.[7]

The most destructive storm on record, Hurricane Andrew, struck Florida in 1992, racking up $17 billion in insured losses and an estimated $30 billion in total losses. Damage from this storm sent seven local insurance companies into bankruptcy. Climate analyst Jeremy Leggett points out that if Hurricane Andrew had struck land 20 miles further north, hitting Miami, total losses could have reached $75 billion. In second place among tropical storms in damages is Typhoon Mireille, which hit Japan in 1991. It caused $5.4 billion in insured damage and left Japan with a total bill of $10 billion.[8]

One of the most destructive floods on record hit China's Yangtze River basin during the summer of 1998. Although insured damage was barely $1 billion, total damage was calculated at $30 billion, and this did not include the indirect costs associated with the resultant economic disruption. This extensive flood, which lasted for several weeks and directly affected 120 million of the 400 million people living in the Yangtze River basin, came close to destabilizing the Chinese economy as riverside factories were forced to close until the floodwaters receded.[9]

The insurance industry is concerned about the effect of global warming on storm intensity. As early as 1990, Munich Re observed, "If water temperatures increase by 0.5 to 1 degrees Celsius in the course of the next few decades, we can expect an extension of the hurricane season by several weeks and a considerable increase in the frequency and intensity of hurricanes. . . . A warmer atmosphere and warmer seas result in greater exchange of energy and add momentum to the vertical exchange processes so crucial to the development of tropical cyclones, tornadoes, thunderstorms, and hailstorms." Trends over the last decade have borne out their concerns. In May 1991 the strongest cyclone on record for the twentieth century wrought death and destruction in Bangladesh, leaving 139,000 dead.[10]

The same increase in temperature that leads to more

frequent, more powerful storms is also raising sea level. Higher temperatures lead to ocean thermal expansion and ice melting. These two trends, both contributing to rising sea level, are creating a new set of challenges that will not be limited to coastal communities. Attempting to cope with rising sea level will place serious financial burdens on coastal countries. Inland communities will be crowded with refugees from areas that are no longer habitable. Food production will be affected. The loss of output as rice-growing river deltas and floodplains are inundated and as local populations expand could lead to a worldwide shortage of rice.

Few countries have researched extensively the effect of rising seas on their economies and population distribution. A World Bank report concludes that a 1-meter sea level rise would inundate half of Bangladesh's riceland. For a country with 133 million people projected to reach 209 million by 2050, the prospect of losing half of its rice harvest is not a pleasant one.[11]

Other Asian countries where rice is grown on low-lying river floodplains include China, India, Indonesia, Myanmar, the Philippines, South Korea, Thailand, and Viet Nam. For Asia, which produces 90 percent of the world's rice and is home to over half its people, rising seas could mean rising rice prices.[12]

Three countries—Thailand, Viet Nam, and the United States—account for two thirds of world rice exports that meet the needs of the 36 rice-importing countries. In these three, only U.S. rice production is relatively immune to the adverse effects of rising seas. The inundation of deltas and river floodplains in Thailand and Viet Nam could easily eliminate their exportable surpluses. Rising sea level has many agricultural consequences, but its potential effect on the world's rice harvest is generating concern about climate change in Asia.[13]

For the United States, which has a coastline of 20,000 kilometers (12,000 miles), a 1-meter rise in sea level—the upper end of the possible projected rise for this century—would inundate 35,000 square kilometers (13,000 square miles). The regions most affected would be the East Coast, from Massachusetts south to Florida, and the Gulf Coast, from Florida to Texas. Except for a few areas, such as the San Francisco Bay area, the West Coast has much steeper, coastal topographic profiles, which limits the damage and disruption from rising seas.[14]

A study from several years ago indicates that even a half-meter rise in sea level would lead to damage and loss of U.S. coastal property totaling $20–150 billion. This figure is rising as the population in coastal counties, now 53 percent of the national total, increases. Unfortunately, growth is most rapid in coastal communities in the south, the region most vulnerable to rising seas and storm surges.[15]

Among the low-lying cities that need to invest heavily in flood defenses in the event of a 1-meter rise in sea level are New Orleans, Miami, Washington, D.C., and New York. Who will bear the cost of building the fortifications against the sea: the cities or the federal government? Should cities begin accumulating funds in advance of potential inundation to ensure that they have the financial wherewithal to build the dams to hold out the sea? If coastal communities are abandoned, are the property owners responsible for dismantling the buildings in the areas that are being inundated? Or can they just leave them as ghost towns that will gradually break up over time as they are battered by the rising sea?

A 1-meter rise in sea level would create millions of refugees. In Bangladesh alone, tens of millions of people would be displaced. Would they move into the already overpopulated interior? Or would they try to migrate to

Europe or to the less densely populated countries like the United States, Canada, Australia, Russia, and Brazil?

What will happen to the major coastal cities that could be partly inundated by a 1-meter rise in sea level, such as Shanghai? Again, will the central government finance the construction of fortifications to protect Shanghai from rising seas and storm surges, or must Shanghai fend for itself? These are questions with no easy answers. Will this generation's legacy to future generations be millions of rising-sea refugees and mega-scale public works projects as coastal communities try to protect themselves from the rising sea level set in motion today?

Intense heat waves can exact a heavy toll in human suffering and even death. A heat wave with temperatures reaching 45 degrees Celsius (113 degrees Fahrenheit) in eastern India in May 2002 took 1,030 lives in the state of Andhra Pradesh alone. Many more died in West Bengal and other neighboring states. To the northwest, in Islamabad, Pakistan, the temperature on June 14 reached a searing 48 degrees Celsius (118 degrees Fahrenheit).[16]

Heat waves can also devastate crops. In the summer of 1988, the United States experienced drought and crop-withering heat simultaneously in its midwestern agricultural heartland. Together they reduced the U.S. grain harvest below domestic consumption for the first time in history.[17]

Fortunately for the more than 100 countries dependent on U.S. grain, the United States was able to satisfy domestic demand by drawing down its vast grain reserves. If such a shortfall were to occur in 2002, the world would be in trouble because the United States no longer has such extensive grain reserves. A similar harvest reduction would translate into reduced exports and in all likelihood a dramatic climb in world grain prices.

Climate change is now a global food security issue.

Higher temperatures mean more extreme climate events. Whether they be droughts, heat waves, storms, or floods, all have the potential of disrupting production and destabilizing grain markets.

One of the difficulties in doing a cost-benefit analysis on the burning of fossil fuels is that those benefiting and those bearing the costs may live on opposite sides of the planet. The United States is the principal source of atmospheric carbon emissions from fossil fuel burning. Bangladesh, a low-lying country, may be one of the principal victims.

The costs and benefits of burning fossil fuels are also separated by time. The benefits from burning fossil fuels are immediate, but the more destructive storms or rising sea level caused by their use may lag by decades, generations, or even centuries. Fortunately, we can now gain the same benefits from more benign energy sources.

Restructuring the Energy Economy

The key to restoring climate stability is shifting from a fossil-fuel-based energy economy to one based on renewable sources of energy and hydrogen. Advancing technologies in the design of wind turbines that have dramatically lowered the cost of wind-generated electricity to the point where it can be used to produce hydrogen from water, along with the evolution of fuel-cell engines, have set the stage for a dramatic restructuring of the world energy economy. The good news is that this shift is under way. The bad news is that it is not happening nearly fast enough to avoid a climate-disrupting buildup in atmospheric CO_2 levels.

The burning of each of the three fossil fuels is now either growing slowly or declining. From 1995 to 2001, the use of oil, the world's leading source of energy, expanded by just over 1 percent a year. Natural gas, the cleanest and least climate-disruptive of the three fossil fuels, grew by less than 3 percent a year.[18]

The burning of coal, the dirtiest and most carbon-intensive fossil fuel, peaked in 1996 and has dropped by 6 percent since then. This historical peaking, marking the first decline in the use of a fossil fuel, may be followed by a similar peaking in oil use within the next 5–15 years.[19]

In contrast, renewables, starting from a small base, are growing at an extraordinary pace. Worldwide, wind electric generation grew by 32 percent a year from 1995 to 2001. (See Table 1–8.) In 2001 alone it grew by a robust 36 percent. And in the United States, wind electric generating capacity jumped by a phenomenal 66 percent in 2001.[20]

Solar cell sales, growing by 21 percent a year from 1995 to 2001, are likely to grow even faster in the years ahead. Once economically competitive only when used in satellites and pocket calculators, solar cells are now becoming competitive for residential lighting in Third World villages not yet connected to the grid. In many countries, if getting electricity to villages means building both a centralized power plant and a grid to deliver the power, it is now often cheaper for families simply to install solar cells. In Andean villages, for example, the monthly installment cost (with a 30-month payment period) on an array of solar cells to provide lighting is comparable to the cost of candles. A similar price relationship exists for the more remote villages in India that depend on kerosene lamps for light.[21]

Another renewable source, one with a largely overlooked potential, is geothermal energy, which is growing at 4 percent a year. This is a vast resource and one that is likely to figure prominently in the energy economies of the Pacific Rim, particularly where widespread volcanic activity indicates that geothermal energy is close to the

Table 1–8. *Trends in Energy Use, by Source, 1995–2001*

Energy Source	Annual Rate of Growth (percent)
Wind power	+32.0
Solar photovoltaics	+21.0
Geothermal power[1]	+ 4.0
Hydroelectric power	+ 0.7
Oil	+ 1.4
Natural Gas	+ 2.6
Nuclear Power	+ 0.3
Coal	– 0.3

[1]Data available through 1999.
Source: See endnote 20.

earth's surface. The western coasts of South America, Central America, and North America have an abundance of geothermal energy. Perhaps the geothermally richest region is the western Pacific, including Indonesia, the Philippines, Japan, and the eastern and southern coasts of China. Another rich region is the Great Rift Valley, which stretches through East Africa up into the Middle East. In fact, the entire eastern Mediterranean is geothermally well endowed. Some countries have enough geothermal energy to meet all their electricity needs.[22]

Hydroelectricity, which supplies over one fifth of the world's electricity, has expanded by 2 percent a year since 1990. In contrast to the other renewable sources of energy, the growth in hydropower is losing momentum as suitable sites for new dams are scarce and as public opposition mounts to large-scale inundation of land, the associated displacement of people, and the disruption of ecosystems.[23]

One of the difficulties in restructuring the energy economy is that doing so typically depends on small, fledgling industries challenging large, well-established, often heavily subsidized industries. One way to accelerate the restructuring needed to stabilize climate is to adopt full-cost pricing, requiring that those using energy pay the full cost of doing so. This approach is discussed further at the end of this section.

Fortuitously, the fastest-growing fossil fuel is natural gas, which is the obvious transition fuel from a carbon-based energy economy to a hydrogen-based one. The natural gas infrastructure, including distribution networks and storage facilities, can easily be adapted for hydrogen as gas reserves are depleted.

As the effects of climate change become clearer, the public's desire to avoid extreme climate events will intensify. As this happens, pressure to raise carbon taxes and reduce income taxes may well rise, providing a strong economic incentive for energy restructuring.

The new century is bringing new directions in the world energy economy. The last century was characterized by the globalization of energy as oil emerged as the leading energy source. Indeed, the entire world became heavily dependent on one region, the Middle East, for a disproportionately large share of its energy. Now as the world turns to wind, solar, and geothermal as the primary energy sources and to hydrogen as an end-use fuel, the energy economy is localizing, reversing the trend of the last hundred years.

Building the Wind-Hydrogen Economy

For many years it appeared that wind would be a cornerstone of the new energy economy, but it now appears that it could become the centerpiece. Between 1995 and 2001, world wind electric generation multiplied nearly fivefold.

(See Figure 1–5.) The generating capacity of 24,000 megawatts at the end of 2001 was sufficient to meet the residential needs of 24 million people at industrial-country consumption levels, a number equal to the combined populations of Denmark, Finland, Norway, and Sweden.[24]

Wind is abundant, cheap, inexhaustible, and clean—four attributes that make it unique. By any yardstick, it is an abundant resource. In the United States, for example, a national wind resource inventory by the Department of Energy reports that the United States is richly endowed with wind energy. The Great Plains, sometimes referred to as the Saudi Arabia of wind energy, could easily supply twice as much electricity as the United States now uses. The United States is not the only country with a wealth of wind. China could double its current electricity generation from wind alone. Europe has enough readily accessible offshore wind energy to satisfy its demand for electricity.[25]

Over the last 15 years, the cost of generating electricity from wind has fallen dramatically, dropping from 38¢ a kilowatt-hour to 4¢ or less at prime wind sites today. Indeed, some recent long-term wind electricity supply contracts have been signed at 3¢ per kilowatt-hour. Wind-generated electricity is now competitive with that generated from other sources, even without including the costs of climate disruption associated with producing electricity from fossil fuels.[26]

Once cheap electricity from wind is available, it can be used to electrolyze water, producing hydrogen. Hydrogen is a way of both storing and efficiently transporting wind energy. Hydrogen is the fuel of choice for the new fuel-cell vehicles that every major automobile manufacturer is working on. Parallel technological advances over the last decade in the design of wind turbines and the evolution of fuel-cell engines have set the stage for a restructuring

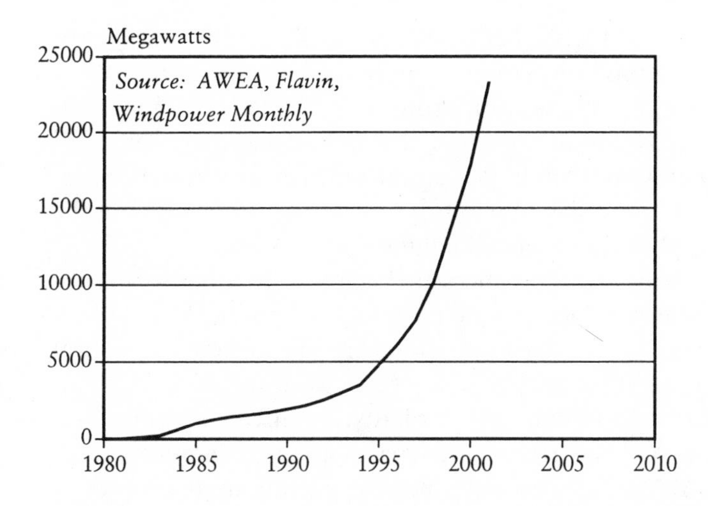

Figure 1–5. *World Wind Energy Generating Capacity, 1980–2001*

of the world energy economy. In the United States, for example, farmers and ranchers, who own most of the wind rights, could one day not only meet most of the country's electricity needs but also supply much of the fuel used in automobiles.

The hydrogen era is beginning to unfold. Both Honda and DaimlerChrysler plan to be on the market with fuel-cell automobiles powered by hydrogen in 2003. Ford expects to be in the market in 2004 with fuel-cell vehicles that run on compressed hydrogen. It will initially concentrate its marketing on fleets of cars that can easily be fueled from a central fueling station.[27]

The Icelandic government is working with a consortium of companies led by Shell and DaimlerChrysler to become the world's first hydrogen-powered economy. The first phase of this program begins in 2003 with the conversion of 3 of Reykjavik's 80 buses from internal combustion to fuel-cell engines. Shell will open its first hydrogen station in Iceland to supply the buses with fuel.[28]

Singapore, anticipating the arrival of fuel-cell-powered automobiles in 2003, has signed a letter of intent with BP to build several hydrogen stations. Air pollution concerns apparently are driving this shift to the non-polluting fuel-cell engine.[29]

Farmers and ranchers are beginning to mobilize in support of developing wind energy. Local political leaders are starting to understand the economic benefits for rural communities of developing their wind resources. In contrast to fossil-fuel power plants, the income from wind-generated electricity tends to stay in the community.

Among the countries moving rapidly to develop their wind resources are Germany, Denmark, Spain, and the United States. Denmark now gets more than 15 percent of its electricity from wind. In Schleswig-Holstein, the northernmost state in Germany, the figure is 28 percent; in the state of Meckenburg-Vorponnen, it is 21 percent. In Navarra, a northern industrial province in Spain, 22 percent of the electricity comes from wind. Wind-generated electricity in California is sufficient to meet the residential needs of San Francisco.[30]

Once a country's wind-generating capacity reaches 100 megawatts, wind resource development tends to accelerate, acquiring a momentum of its own. At this point, countries appear to have the engineering experience, the financial structures, and the professional expertise in place to develop this vast but largely untapped resource. Some 16 countries, home to half the world's people, are now in this group.

In the wind-hydrogen economy, wind turbines will replace smokestacks. Hydrogen generators will replace oil refineries. Fuel cells will replace internal combustion engines. Fuel cells run on hydrogen do not produce any pollutants nor do they make any noise. Quiet cities with clean air will replace noisy, polluted cities.

Among the countries setting ambitious wind development goals are Germany, Spain, the United Kingdom, France, Argentina, India, China, Brazil, and the United States. A 3,000-megawatt wind farm in the early planning stages in South Dakota is one of the largest energy projects being contemplated anywhere. Some wind-rich countries could become hydrogen exporters in the new energy economy: Canada, which is richly endowed with wind but sparsely populated, and Argentina, with world-class wind resources in Patagonia, could become leading hydrogen exporters. Both countries could generate vast quantities of hydrogen, supplying it in perpetuity, to more densely populated countries with less favorable wind/population ratios.[31]

In the northwestern United States, where hydropower is already well established, cities such as Salem, Oregon, are now moving to get the rest of their electricity from wind, making them entirely independent of fossil fuels for electricity. The shift from a fossil-fuel-based economy to a renewable-energy-based one is under way. The challenge now is to accelerate that transition before climate change spirals out of control.[32]

Fixing the Market

The market is a remarkable institution. It allocates scarce resources with an efficiency that no central planning body can match. It easily balances supply and demand and it sets prices. As noted earlier, however, it does have three fundamental weaknesses, namely its failure to incorporate the indirect costs of providing goods or services into prices, its inability to value nature's services properly, and its lack of respect for the sustainable-yield thresholds of natural systems such as fisheries, forests, rangelands, and aquifers.

Throughout most of recorded history, there was little reason to be concerned about the sustainable yields of natural systems, the value of nature's services, or the indirect costs of economic activity because they were rarely an issue. But with the 19-fold expansion in the world economy over the last century, the failure to address these market shortcomings will be costly.[33]

As the global economy has expanded and as technology has evolved, the indirect costs of some goods and services have become far larger than the price fixed by the market. As noted earlier, the price of a gallon of gasoline includes the cost of production but not the expense of treating respiratory illnesses from breathing polluted air or the repair bill from acid rain damage. Nor does it cover the cost of rising global temperature, of more destructive storms, and of relocating future millions of rising-sea refugees. As the market is now organized, the motorist burning the gasoline does not bear the cost of rising sea level and the potential losses of ocean-front property, the evacuation of coastal cities, or the loss of the rice harvest from the inundation of low-lying river deltas and floodplains.

A recent study from the Centers for Disease Control and Prevention, a U.S. government agency, on the indirect costs of smoking cigarettes illustrates the kind of analysis needed for burning fossil fuels. The study reports that the market price of a pack of cigarettes, which includes the cost of growing the tobacco and processing it into cigarettes, is $2.80 in the United States. But the cost to society of smoking a pack of cigarettes, including both the medical costs of treating smoking-related illnesses and the cost of lost worker productivity as a result of absenteeism, is $7.18. The issue is not whether the additional $7.18 is paid—clearly it is paid by someone, either the smoker, the smoker's employer, or taxpayers. The full cost of each pack of cigarettes smoked is thus $9.98.[34]

If the indirect costs of burning a gallon of gasoline were incorporated into its price, would the $1.60 that Americans typically pay be $4, $6, or maybe $10 per gallon? Would the cost to society of burning a gallon of gasoline be more or less than the $7.18 from smoking a pack of cigarettes? Intelligent investment and purchasing decisions for society, whether by government policymakers, corporate planners, or individual consumers, depend on estimating these costs and incorporating them into the price of the gasoline. If the unlevied cost of using a product, such as cigarettes or gasoline, leads to extreme market distortions, it could eventually lead to economic bankruptcy.

Another market shortcoming is its failure to price properly the many services that nature provides, which are often taken for granted. Nature converts salt water from the oceans into fresh water through evaporation. It pollinates crops, recycles nutrients, and purifies water. The destruction of natural systems deprives society of these services—services with a value that society is only beginning to recognize.

New York City, with its population of nearly 17 million, recently discovered just how valuable nature's water purification service is. Faced with the residential and industrial development of the Catskill forest region and the associated pollution of water in the watershed that is the source of New York's water, the city was told it needed a water purification plant that would cost $8 billion to build and $300 million a year to operate. The bill for this would reach $11 billion over 10 years. City officials realized that they could restore the watershed to its natural condition for only $2 billion, and let nature purify the water, thus avoiding the need for the purification plant and saving taxpayers $9 billion.[35]

The Chinese have learned the hard way the value of the flood control services provided by forests. During the summer of 1998, several weeks of some of the worst flooding on record in the Yangtze River basin wreaked enormous havoc. Direct damage caused by the flood totaled some $30 billion, according to Munich Re. Some 120 million people were directly affected by the flooding. Up until mid-August the Chinese government had been describing the catastrophe as an act of nature. But they finally recognized that deforestation, specifically the loss of 85 percent of the original forest cover in the Yangtze River basin, was a major contributor to the flooding. Once they understood this, they banned tree cutting in the basin. They justified the ban by noting that trees standing are worth three times as much as trees cut. Chinese leaders were acknowledging that the flood control value of the forests in the Yangtze River basin was worth three times as much as the lumber in the trees. The market values lumber in the trees, but not the flood control service provided by the forests. In addition to the ban on tree cutting, the government launched a tree planting program to restore the flood control service.[36]

The $30 billion worth of damage from the Yangtze flood, plus the disruption it caused in the heavily industrialized Yangtze river basin, nearly derailed the Chinese economy. To put the loss in perspective, China is the world's leading producer of both wheat and rice, but the $30 billion of flood damage exceeds the value of the annual wheat and rice harvests combined. It was the recognition in Beijing of the scale of the damage and disruption from the flooding that led to the abrupt shift in policy from tree cutting to tree planting.[37]

The market's lack of appreciation of the value of nature's services is pervasive. In the two cases just cited, the governments of New York City and China made major policy changes once they recognized the value of nature's services, values that were not reflected in the market. Will the world one day reach a similar judgment on the costs of climate change?

Another area in which the market is inept is in recognizing the sustainable-yield thresholds of natural systems. If an oceanic fishery's catch increases over time in response to demand until it exceeds the fishery's sustainable yield and stocks begin to decline, then fish prices will begin to rise. The market's response to higher prices is to invest more in fishing trawlers, a response that ensures collapse of the fishery. Further exacerbating this problem are the subsidies that governments pay the fishing industries, which distort the market.

What some governments are realizing is that protecting an economy's natural resource base depends on introducing the concept of sustainable yield into the market place. Australia, concerned in 1986 about the overfishing of its lobster fishery, estimated the sustainable yield of the fishery and then issued fishing permits totaling that amount. Fishers could then bid for the permits. In effect, the government decided how many lobsters could be taken each year on a sustainable basis and then let the market decide how much the permits were worth. Once the permit trading system was adopted, the fishery stabilized and has operated on a sustainable basis ever since.[38]

A similar situation exists with aquifers. As the demand for water increases, the pumping eventually exceeds the sustainable yield of the aquifer and the water table starts to fall. The market simply says, "drill deeper." But all this does is allow the use of water to continue to increase while the aquifer is being depleted. Once that happens, the rate of pumping is necessarily reduced to the rate of recharge. But if the level of use at this point is double the rate of recharge, which can easily be the case, then water use is abruptly cut in half. Needless to say, the adjustments to aquifer depletion can be abrupt and destabilizing.

Enlightened government policy could intervene by establishing the sustainable yield of the aquifer and auctioning off the rights to pump that amount of water. This lets the market allocate the water to the more high-valued uses, while stabilizing the water table.

There are two dimensions of overpumping that are of concern. One, once the rising level of water use exceeds the sustainable yield of the aquifer, the gap between that use and the sustainable yield widens each year until the aquifer is dry. This means that if countries delay until depletion occurs, they will face wrenching reductions in the use of water. Two, the overpumping of aquifers is proceeding simultaneously in literally scores of countries, which means that at some point in the not-too-distant future the world will face simultaneous "water shocks" as aquifer depletion forces abrupt and, in many cases, substantial reductions in water use.

Unless governments are prepared to intervene in the market to get prices to tell the ecological truth, to value nature's services, and to respect the sustainable yields of natural systems, then the economy will eventually destroy its natural support systems. Thus far the ecological and economic consequences of market distortions in our modern civilization have been mostly local and manageable. But if they continue to increase, they will eventually become worldwide, setting the stage for global economic decline.

2

Eco-Economy Indicators: Twelve Trends to Track

The dozen indicators in this section were chosen to measure progress, or the lack thereof, in building an eco-economy—one that respects the principles of ecology.

Population is selected because although it is a social indicator, it is also a basic environmental indicator. During most of the past 4 million years, our existence as a species was precarious, our numbers small. Now we are so numerous and leave such a large ecological footprint that we threaten the existence of the millions of other species with whom we share the planet. When assessing the adequacy of basic resources such as land and water over time, population size is the universal denominator, always shrinking per capita availability as it expands.

Economic growth is included because, given the way the world now does business, the size of the economy is the best single measure of the mounting pressure on the earth's environment. It combines the effects of both population growth and rising individual consumption to give us a sense of how much the pressure is increasing.

The third trend, grain production, is the best indicator of the adequacy of the food supply. On average, half of all the calories we consume come directly from grain and a large part of the remainder come from the indirect consumption of grain in the form of meat, milk, eggs, and farmed fish. Grain production is a useful indicator of growing food demand in that increased output reflects population growth and also rising affluence, with its associated rise in consumption of grain-fed livestock products.

The world fish catch is a useful measure of the productivity and health of the oceanic ecosystem that covers 70 percent of the earth's surface. The extent to which world demand for seafood is outrunning the sustainable yield of fisheries can be seen in shrinking fishery stocks, declining catches, and collapsing fisheries.

Forest cover is one of the best single indicators of changes in land use. Shrinking forest cover shows we are cutting more trees than we are planting. The shrinkage of forested area means not only that the forest's capacity to supply products is diminished, but also that its capacity to provide services, such as flood control, soil protection, and the purification of water, is also reduced.

Water scarcity may be the most underrated resource issue the world is facing today. Because water was relatively abundant throughout most of our existence, we came to take it for granted. Now we see that water tables are falling in scores of countries. The data show that these individual countries and indeed the entire world soon will be facing "water shocks" as aquifers are depleted and the water supply is abruptly reduced.

Carbon emissions are revealing because as the atmospheric concentration of carbon dioxide changes, so does the earth's temperature. Thus carbon emissions tell us a lot about ourselves and our current habits and provide clues about the kind of world we will be leaving for future generations. Will we be leaving them a stable climate, or will it be a world of searing heat waves, more destructive storms, melting glaciers, and rising sea level—a world besieged by millions of rising-sea refugees?

Just as taking our own body temperature is one of the best measures of our health and well-being, so temperature is also a measure of how well we are taking care of the earth, the only planet known to support life. For the first time in human history, our actions are linked to changes in the global temperature. Who would have thought a generation ago that the thermometer might become the device with which we assessed the human prospect?

Ice melting is included as one of our indicators because it is both one of the most sensitive and one of the most visible effects of rising temperature. There are many other indicators of rising temperatures, such as forests beginning to migrate, tropical diseases moving into higher latitudes, or tree lines moving upward on mountains, but none are quite so visible and perhaps disturbing as the melting of glaciers and ice sheets. Since so much of the world's water is stored in ice on land, its melting raises sea level, directly influencing the human prospect.

Wind electric generating capacity is included here not because of its importance as an energy source today but because of its likely importance tomorrow. Advances in wind turbine design have set the stage for wind power to become the foundation of the new energy economy. Because it is abundant, cheap, inexhaustible, and clean, wind energy is now growing by leaps and bounds. Examining the rate at which wind generating capacity is expanding compared with fossil fuels gives us a sense of how fast the eco-economy is unfolding.

Bicycles are included because their annual sales are more than double those of automobiles. Their sales also measure our ability to reduce traffic congestion, lower air pollution, increase mobility, and provide exercise—a counter to the obesity that is now engulfing urban populations everywhere.

Solar cells are a trend to track because of their likely importance as a future source of energy. On the falling cost curve, solar cells are several years behind wind. Solar cell sales in 2001 of nearly 400 megawatts of generating capacity represent by far the largest annual sales to date, but still this is the equivalent of the output of only a single power plant. The promise lies in the future, where—as it continues to fall—the cost will cross a critical threshold where production will begin to jump. At least one major manufacturer is planning a doubling of production this year.

Population Growing by 80 Million Annually

Janet Larsen

World population climbed to 6.2 billion in 2002, up almost 80 million or 1.3 percent from 2001. Population growth rates soared following World War II as health care improved and death rates fell. After peaking at 2.1 percent around 1970, annual world population growth fell to 1.3 percent by 1999. But even while global growth is slowing, there is a large disparity among the growth rates of individual nations, and human numbers overall continue to climb.[1]

For at least 25 years, 20 European countries and Japan have had below replacement-level fertility rates (2.1 children per woman). By now a total of 44 countries have fertility levels that low. Without the projected gain of 2 million immigrants a year from developing countries, many industrial nations would shortly experience population declines.[2]

In much of the developing world, however—home to nearly 5 billion people—populations are still growing rapidly. Even with anticipated declines in fertility rates, the developing world is projected to have 8.2 billion people by 2050. Six countries account for half of the world's annual addition: India (16 million), China (9 million), Pakistan (4 million), Nigeria (4 million), Bangladesh (3 million), and Indonesia (2 million).[3]

The 48 countries classified as least developed have even more rapid population growth. If current trends continue, the combined populations of these nations will almost triple by mid-century—from 658 million to 1.8 billion. Among the 16 countries with extremely high fertility rates (seven children or more per woman) are Afghanistan, Angola, Burkina Faso, Burundi, Liberia, Mali, Niger, Somalia, Uganda, and Yemen.[4]

Fertility rates in countries at the intermediate level, where women have between 2.1 and 5 children on average, are expected to drop below replacement level by 2050. This group includes India, Pakistan, South Korea, and Egypt, which were among the first to realize that rapid population growth makes it difficult to reach socioeconomic goals. With high population densities in their most fertile land areas, these countries recognized that fast-growing populations test the limits of both social services and nature's services.[5]

Although the availability of effective contraception is a key to slowing population growth, some 350 million women worldwide still lack access to family planning. Filling the unmet need for family planning could reduce population growth by as much as a third, given the estimated number of unintended pregnancies in the developing world.[6]

Fertility inversely correlates with levels of female education and employment. The more schooling women have, the fewer children they bear. Educating women and men about family planning services and making such services readily and discreetly available could profoundly reduce future world population size and poverty. Government-supported family planning programs increase access to reproductive and general health care. High per capita incomes, low child mortality, urbanization, and industrialization also can play a role in lowering fertility.[7]

At the International Conference on Population and Development held in Cairo in 1994, parties agreed to fund a 20-year population and reproductive health program, with developing countries covering two thirds of the bill and donor countries paying the rest. The total yearly spending was expected to be $17 billion until 2000, and then climb to $22 billion by 2015. While developing countries have largely honored their commitment, donor countries have contributed only one third of their allotted share. The results of this shortfall are that training and services have not expanded as promised, which researchers calculate meant that between 1994 and 2000 some 122 million women became pregnant unintentionally. One third of them had abortions. In addition, an estimated 65,000 unintentionally pregnant women died in childbirth and 844,000 suffered chronic or permanent injury as a result of their pregnancies.[8]

Epidemics like HIV/AIDS reduce population projections by increasing morbidity and mortality and also by lowering fertility. AIDS is altering the demographics of many countries, especially in Africa. In Botswana, 36 percent of the adult population is HIV-positive. There, life expectancy has fallen precipitously from 70 years to 36, and Botswana's total population in 2015 is projected to be 28 percent smaller than it would be in the absence of AIDS. In Zimbabwe, life expectancy has dropped to 43 years, and in South Africa, to 47.[9]

Today nearly half the world's people live in cities, where concentrated populations facilitate disease transmission. Fortunately, high population densities also enable potentially efficient provision of services such as health care and education, if there is the political and community will.[10]

Urban areas are expected to absorb almost all of the population growth of the next 30 years. After centuries of rural-to-urban migration, three fourths of people in the industrial world live in cities. Developing countries are following this same pattern. In 1950, 18 percent of people in the developing world were urban dwellers. This more than doubled to 40 percent in 2000, and is projected to reach 56 percent by 2030, when 60 percent of the world will live in cities.[11]

Almost one third of the world today is under the age of 14. History's largest generation of young people is reaching or will soon reach reproductive age, intensifying population momentum. As medical advances allow people to live longer than ever before, the global population is also aging. Today more than 606 million people are older than 60—a number due to reach 2 billion by 2050.[12]

The gap between the U.N. high-growth projections for 2050 of 10.9 billion and the low-end scenario of 7.9 billion is equal to about half the world's current population. (See Figure 2–1.) With water and land in limited supply worldwide, whether the world moves to the higher or lower number may have more influence on environmental and social sustainability than any other variable.[13]

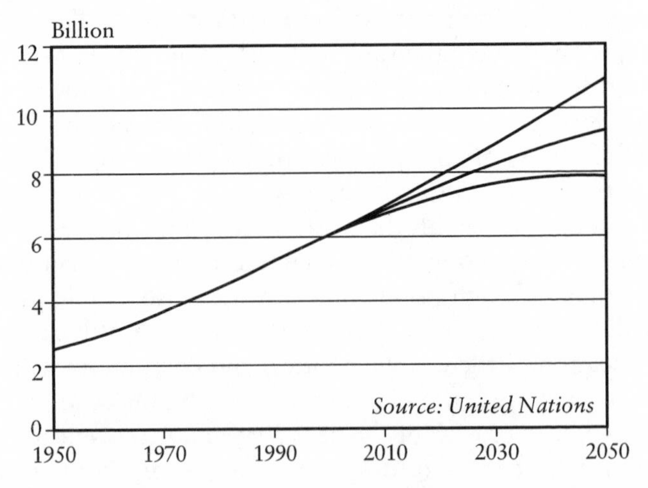

Figure 2–1. *World Population, 1950–2001, with Projections to 2050, Under Three Assumptions of Growth*

Economic Growth Losing Momentum

Lester R. Brown

In 2001, the global economy expanded by scarcely 2 percent, the slowest rate in many years. (See Figure 2–2.) With growth in the economy barely exceeding that of population, gross domestic product per person climbed from $7,392 to $7,454, a gain of less than 1 percent.[1]

Much of the global slowdown was attributable to the United States—both because it is far and away the world's largest economy and because it is the principal export market for so many countries. After expanding by a robust 4.1 percent in 2000, the U.S. economy grew by only 1.2 percent in 2001. Meanwhile, the Canadian economy was slowing in sync, dropping from an increase of 4.4 percent to 1.5 percent.[2]

In Western Europe, the four large industrial countries all experienced declines in growth in 2001. France, Italy, and the United Kingdom each dropped from 3 percent or better to around 2 percent, while Germany—the largest of the four—fell from 3 percent to less than 1.[3]

Growth in Latin America's large economies also slowed substantially. Brazil, the region's biggest, dropped from 4.4-percent growth to 1.5 percent. Argentina, in serious difficulty in recent years as a result of the cumulative effects of economic mismanagement, saw its economy shrink by nearly 5 percent in 2001. (Worse is yet to come there in 2002.) In Mexico, the other large economy in the region, growth dropped from 6.6 percent in 2000 to

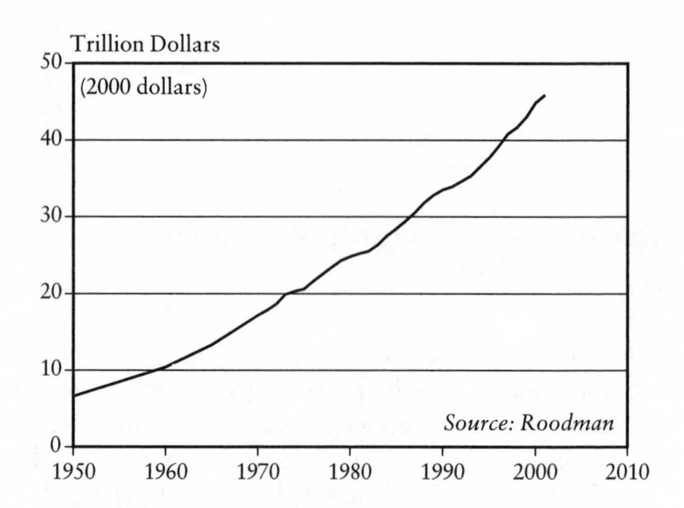

Figure 2–2. *Gross World Product, 1950–2001*

zero in 2001—one of the biggest drops recorded among the world's larger economies.[4]

Economic growth was also losing momentum in the Middle East. For Saudi Arabia, the world's leading oil exporter, it dropped from 4.5 percent in 2000 to 2.2 percent in 2001. For Iran, growth remained steady at 5 percent in 2000 and 2001. Egypt, meanwhile, dropped from 5 percent to just over 3 percent.[5]

In Asia, Japan continued to perform poorly, going from a modest growth of 2.2 percent in 2000 to an actual decline in 2001 (at –0.4 percent). Until it considers major economic reforms, including clearing up a dangerously heavy load of bad bank loans, Japan will have trouble sustaining economic growth. South Korea, which had achieved 9-percent growth in 2000, saw its expansion drop to 3 percent in 2001.[6]

The developing countries of Southeast Asia, meanwhile, did not fare well. Their overall growth rate declined. Indonesia dropped from nearly 5 percent to 3 percent. Thailand dropped from 4.6 percent to 1.8. Malaysia, however, took a bigger hit, dropping from 8-percent growth to nearly zero.[7]

In 2001, India's growth dropped from 5.4 to 4.3 percent and Bangladesh's from 5.5 to 4.5 percent. Pakistan was stable, with an expansion of just under 4 percent in both 2000 and 2001.[8]

China continued as the star economic performer of the region, expanding by 8 percent in 2000 and dropping only slightly to 7.3 percent in 2001. Questions remain, however, about economic accounting in China, with several indirect indicators suggesting that growth has been consistently overstated.[9]

In the former Soviet republics, Russia's economic growth dropped from 9 percent in 2000 to 5 percent in 2001. In contrast, the Ukrainian economy, which has been struggling for many years, went against the tide—expanding from 6-percent growth in 2000 to 9 percent in 2001.[10]

In Africa, a few countries also countered the global trend of slower growth. Algeria's economy climbed from 2.4-percent growth in 2000 to 3.5 percent in 2001. Morocco grew from 2.4 percent to 6.3 percent. And Nigeria, helped by rising oil prices, held steady at 4 percent.[11]

Economic and social progress have not come easily in Africa. Although the region's economy resumed growth during the last decade, it was not able to match the increase in population. As a result, income per person in sub-Saharan Africa declined by some 12 percent from 1980 to 1999. Life expectancy, perhaps the best social indicator of progress, is only 50 years, and that may fall during this decade as the HIV epidemic shortens millions of lives.[12]

What these data on economic growth from the International Monetary Fund do not show is the share of economic output that is environmentally unsustainable. Available evidence suggests that as much as 8 percent of the world grain harvest may be based on the unsustainable use of water. At some point, overpumping of water will come to a halt either because it is too costly to pump from a continually falling water table or, perhaps more likely, because the aquifer is depleted. If it is a rechargeable aquifer, depletion means that pumping will necessarily be reduced to the rate of recharge. If it is a fossil aquifer—that is, nonrechargeable—pumping ends.[13]

A similar situation exists for forest products, where clearcutting and the shrinkage of the remaining forested area are reducing the long-term yield of the earth's forests. Deforestation may initially have a positive effect, but it brings its own set of costs in soil erosion and flooding.

Fisheries, too, are being overharvested in order to maximize short-term income. Some three fourths of oceanic fisheries are being fished at or beyond their sustainable yield. In some cases, governments are cutting back on the catch to try and save the fisheries. In others, the fisheries are simply collapsing. The result is the same: a reduced overall catch.[14]

These and many other trends simply underline the risks associated with dependence on economic data that do not distinguish between sustainable and unsustainable output. The failure to do so is leading to an exaggerated sense of progress and to a false sense of security.

Grain Harvest Growth Slowing

Lester R. Brown

The 2001 world grain harvest of 1,853 million tons was up 1 percent from the 2000 harvest, but below the all-time high of 1,880 million tons in 1997. (See Figure 2–3.) The U.S. Department of Agriculture reports that the harvest in 2001 fell 40 million tons short of estimated consumption. This comes on the heels of a poor crop in 2000, when output was 36 million tons short.[1]

These two consecutive disappointing harvests have reduced this year's projected world carryover stocks of grain, the amount in the bin when the new harvest begins, to 24 percent of annual consumption, the lowest level in 20 years. With stocks at such a low level, all eyes will be on the harvest in 2002. Another shortfall could lead to rising grain prices and higher prices for bread, meat, milk, eggs, and other products derived directly or indirectly from grain.[2]

The poor harvests of the last two years were largely due to weak grain prices, drought, and spreading water shortages. Grain prices among the lowest in two decades have discouraged farmers from investing in production-boosting measures.[3]

Prices that are too low to stimulate adequate production can be quickly remedied as the market responds to tighter supplies. But dealing with the water shortages that result from drought, aquifer depletion, and the diversion of scarce water to cities is much more difficult.

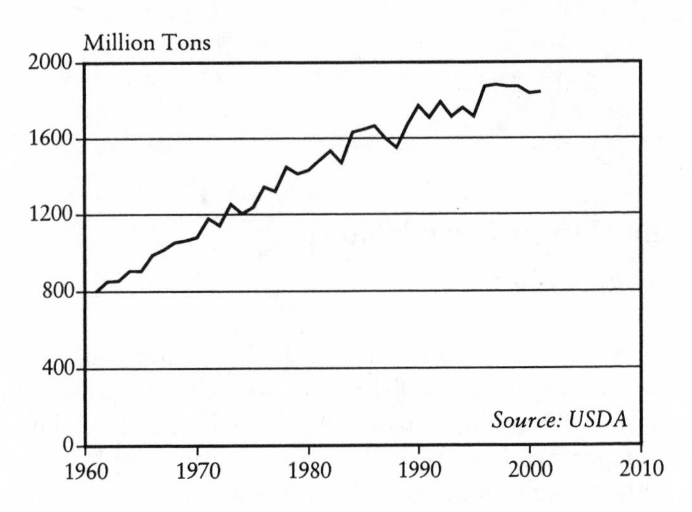

Figure 2–3. *World Grain Production, 1961–2001*

Water tables are now falling in key food-producing regions—the North China Plain, the Punjab in India, and the southern Great Plains of the United States. The North China Plain accounts for a quarter of China's grain harvest. The Punjab, a highly productive piece of agricultural real estate, is India's breadbasket. And the southern Great Plains helps make the United States the world's leading wheat exporter.[4]

In an increasingly integrated world economy, water shortages are crossing national boundaries via the international grain trade. Since it takes 1,000 tons of water to produce 1 ton of grain, the most efficient way for water-deficit countries to import water is to buy grain from elsewhere.[5]

The fastest-growing grain import market in the world today is North Africa and the Middle East, the region with the most serious water shortages. Virtually every country in this region—stretching from Morocco across the northern tier of Africa and the Middle East through Iran—is facing water shortages. With supplies limited, countries satisfy the growing demand for water in cities and industry by taking it from agriculture. Then they import grain to offset the loss of production capacity.[6]

In recent years, grain imports into Iran, a water-short, grain-deficit country, have eclipsed those of Japan, long the world's leading wheat importer. Last year, Egypt also moved ahead of Japan. Both Iran and Egypt now import over 40 percent of the grain they consume. The populations of both countries are continuing to grow, but their water supplies are not.[7]

Grain exporters are, in effect, water exporters. Canada, where water exports are a politically sensitive issue, is one of the world's leading exporters of water in the form of grain. The 18 million tons of grain, mostly wheat, that it ships abroad each year embody 18 billion tons of water. Similarly, U.S. annual grain exports of 90 million tons of grain represent 90 billion tons of water, an amount that exceeds the annual flow of the Missouri River.[8]

The adequacy of food and water supplies are closely linked. Some 70 percent of all water that is pumped from underground or diverted from rivers is used to produce food, while 20 percent is used by industry and 10 percent goes to residential uses. With 60 percent of the world's grain harvest produced on irrigated land, anything that reduces the irrigation water supply reduces the food supply.[9]

The wild card in the world grain market is China. It accounted for virtually all of the world grain harvest shortfalls in 2000 and 2001. Indeed, in two years, it has reduced grain stocks by nearly 80 million tons.[10]

Among the forces shrinking China's grain harvest are severe drought in northern China during the last two years, spreading irrigation-water shortages as aquifers are depleted and as water is diverted to cities, and a lowering of support prices. The drought will eventually end, but water shortages will not. In a country dependent on irrigated land for 70 percent of its grain, water shortages are fast becoming a security issue.[11]

In 1994, in an ambitious and successful effort to be self-sufficient, China raised grain support prices by 40 percent. Unfortunately the drain on the treasury was too great, so the support prices were eventually lowered, dropping close to world market levels.[12]

China has absorbed the harvest shortfall of the last two years by drawing down stocks, but there are signs that supplies are now tightening. If this huge nation has another large harvest shortfall, it will likely have to import substantial quantities of grain to maintain food price stability.[13]

If the 2002 world grain harvest falls short of consumption when stocks are at a near-record low, prices will rise. Higher prices will curb demand, particularly the feeding of grain to livestock, and will encourage production. Supply and demand will again be in balance, but at a higher price.

If world grain demand continues to grow during this coming year at the 16-million-ton-per-year pace of the last decade, then the 2002 harvest will have to jump by 70 million tons to avoid a further drawdown in stocks. Whether this can occur, in the face of spreading water shortages, remains to be seen. The new reality is that if the world is facing water shortages, it is also facing food shortages.

A review of the demographic map reveals another troubling reality. Most of the 80 million people added to world population each year live in countries that already have water shortages. Restoring a balance between water supply and needs worldwide may now depend on stabilizing population in water-deficit countries.[14]

Fish Catch Leveling Off

Janet Larsen

The world fish catch in 2000, the last year for which global data are available, was reported at 94.8 million tons. After decades of steady growth, the oceanic fish catch has plateaued and since the late 1980s has fluctuated between 85 million and 95 million tons. (See Figure 2–4.) Some three fourths of oceanic fisheries are fished at or beyond their sustainable yields. In one third of these, stocks are declining.[1]

Some scientists, when correcting for suspected overreporting by China, the world's leading fishing nation, believe that global catch has actually declined by 360,000 tons each year since 1988. When catch of the highly variable stocks of Peruvian anchovetas, a species substantially affected by El Niño/Southern Oscillation events, is excluded, the world fish catch appears to have declined by 660,000 tons a year during that time.[2]

Recent evidence points to a rapid decline in production of the North Atlantic Ocean, where catches of many popular fish species, including cod, tuna, haddock, flounder, and hake, have dropped by half within the past 50 years, even though fishing efforts tripled. Previous infamous collapses, like that of the Newfoundland cod fishery, were local in scale, but this decline is ocean-wide.[3]

At least $2.5 billion of government money goes to subsidize fishing in the North Atlantic each year, supporting incomes and paying portions of boat fuel and

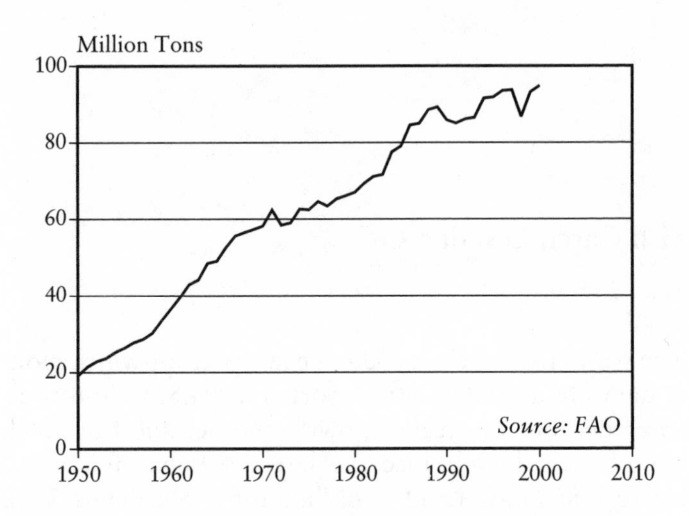

Figure 2–4. *World Fish Catch, 1950–2000*

equipment bills. Worldwide, fishing subsidies total at least $15 billion, but may be substantially higher. In 1993, the U.N. Food and Agriculture Organization reported that the operating costs of fisheries around the world exceeded commercial revenues by over $50 billion each year. Without subsidies, the world's fishing industry would be bankrupt.[4]

About 950 million people worldwide rely on fish as their primary source of protein. In addition, ocean fisheries and fish-related industries sustain the livelihoods of some 200 million people. These are high numbers to sustain on a bankrupt industry.[5]

Subsidies hide the fact that current fishing practices are unsustainable, both economically and ecologically. Subsidy money has helped to build a technologically advanced global fishing fleet of over 23,000 ships weighing more than 100 tons each. Massive ships, such as trawlers, drag big nets—quickly catching large quantities of fish and bycatch. Some vessels have onboard processing facilities. Large ships consume a great deal of energy: it takes twice as much fuel to capture a ton of fish today as it did 20 years ago. Overall, the world's fishing fleet has the capacity to catch fish at more than twice the fisheries' sustainable yields.[6]

As fish harvests from the ocean are steady or declining, production of fish from farms (aquaculture) is booming. Since 1990, aquaculture production has grown by almost 10 percent each year, more than twice the rate for poultry, the second fastest-growing sector of the animal protein economy. Total fish-farm production in 2000 was almost 36 million tons. In 1950, aquaculture provided less than 1 percent of the fish supply; now it accounts for a full 27 percent of the world fish market.[7]

Growing fish in pens and ponds could reduce pressure on oceanic fisheries, but only if it is done wisely. A number of popular farmed fish, like salmon and shrimp, are carnivorous, requiring fish from the oceans to be harvested to provide fish meal and fish oil for their food. Some species require up to 5 kilograms of wild fish for each kilogram of fish produced. Harvesting fish for feed can empty oceans of smaller fish, depriving larger wild fish of their food supply.[8]

China, which provides 23 million tons of the world aquaculture output, has farmed fish for thousands of years. It now devotes some 5 million hectares of land to farming primarily herbivorous fish. An additional 1.7 million hectares of rice paddies double as fish ponds. China has developed an innovative carp polyculture, in which several carp species with complementary feeding habits are grown together as they would in natural ecosystems.[9]

China's onshore, integrated aquaculture and agriculture production system can serve as a model for aquaculturalists. Onshore production can minimize problems that plague marine aquaculture operations, such as coastal habitat destruction and excessive nutrient pollution, which can cause algal blooms. It also reduces the risk of introducing nonnative species through escapes and spreading diseases that fish in high-density confinement are prone to.[10]

For a number of oceanic fisheries, a deliberate reduction of fishing, along with the development of "no-take" protected areas, is the only way for stocks to rebuild. Marine reserves have been shown to increase fish populations and diversity and to produce larger fish both within their boundaries as well as in commercially accessible waters. In a matter of a few years, a nearby off-limits area can revive a foundering fishery.[11]

To protect wild stocks, consumers can reduce their overall fish consumption, or at least purchase responsibly produced herbivorous fish or those caught from well-managed fisheries. The Marine Stewardship Council, an independently operated international accreditation organization, has certified six fisheries as sustainable. Careful management of fisheries can be likened to prudent use of an endowment: if the principal, or the stock, is conserved, people can live off the interest indefinitely.[12]

Forest Cover Shrinking

Janet Larsen

Global forest cover is a key indicator of the health of the planet. An intact forest cycles nutrients, regulates climate, stabilizes soil, treats waste, provides habitat, and offers opportunities for recreation. By a conservative tally, these services are worth more than $4.7 trillion, a total equal to one tenth of the gross world product. Forests also supply goods, including food, medicines, and a large array of wood-based products.[1]

Forests worldwide cover some 3.9 billion hectares—almost a third of the earth's land surface excluding Antarctica and Greenland. Though vast, this wooded area is only half the size of forested land at the dawn of agriculture some 11,000 years ago. Most forests are no longer in their original condition, having changed in composition and quality.[2]

Global estimates of forest cover change are difficult to make because of conflicting definitions of what constitutes a forest, lack of satellite and radar data, and unmonitored land use change. The U.N. Food and Agriculture Organization conservatively estimates that the world lost 94 million hectares of forest in the last decade of the twentieth century. (See Table 2–1.) This number assumes that developing countries lost 130 million hectares while the industrial world gained 36 million hectares as abandoned agricultural areas returned to forest. The yearly loss of natural forests during this period, which includes defor-

Table 2–1. *Change in World Forest Cover, 1990–2000*

Continent	Total Forest 1990	Total Forest 2000	Change, 1990–2000
	(million hectares)		(percent)
Africa	702	650	–7.8
Asia	551	548	–7.0
Oceania	201	198	–1.8
Europe	1,030	1,039	+0.8
North and Central America	555	549	–1.0
South America	923	886	–4.1
World	3,963	3,869	–2.2

Source: U.N. Food and Agriculture Organization, *State of the World's Forests 2001* (Rome: 2001).

estation plus the conversion of natural forests to tree plantations, was 16 million hectares—94 percent of which occurred in the tropics.[3]

During the 1990s, Brazil suffered the heaviest loss of forest—23 million hectares. South America as a whole saw net losses of 37 million hectares. In Africa, 52 million hectares were destroyed. Sudan, Zambia, and the Democratic Republic of the Congo account for half of Africa's forest loss. While the United States gained 4 million hectares of forests, Mexico lost over 6 million, although government reports reveal the loss may be even higher. The total net losses for North and Central America were 6 million hectares.[4]

A massive reforestation campaign in China meant the country added an average of 1.8 million hectares each year during this period, largely because bans on deforestation near the end of the decade heightened the country's reliance on plantations and imports of forest products from other nations. In Indonesia, where tree felling destroyed 13 million hectares over the decade, forest loss has accelerated and now averages 2 million hectares each year. Over the decade, forest cover in all of Asia declined by 4 million hectares.[5]

Although FAO data suggest that world forest loss is slowing, deforestation in tropical areas is accelerating, likely exceeding 13 million hectares each year. As tree cutting in many parts of the world accelerates, nearly half of the remaining forests are at risk. The World Resources Institute estimates that about 40 percent of the world's intact forests will be gone within 10–20 years, if not sooner, considering current deforestation rates.[6]

Wood consumption drives deforestation. Since 1960, global industrial wood production has risen by 50 percent, to 1.5 billion cubic meters, four fifths of which is from primary and secondary-growth forests. About the same quantity, 1.8 billion cubic meters, is burned directly as wood fuel each year in developing countries.[7]

Worldwide, only some 290 million hectares of forested land are under protection from logging, but even protected areas are threatened by illegal exploitation. Of 200 areas of high biological diversity worldwide, illegal logging threatens 65 percent. All told, illegal logging has devastated public forests around the globe, reducing incentives for locals to invest in sustainable forestry and accumulating losses of revenue to governments of some $15 billion annually.[8]

Forest plantations now cover more than 187 million hectares, less than 5 percent of total forested area, but account for 20 percent of current world wood production. As natural forests are exhausted or come under protection, a growing share of future wood demand will be satisfied from tree farms.[9]

Well-planned and managed plantations can efficiently satisfy timber demand. Unfortunately, the world has seen many plantations raised at the expense of old growth or other extremely diverse natural forests. In some cases, governments grant forest concessions to logging companies contingent on their planting of replacement trees, but after the companies clearcut, they leave the land bare and move to new areas. In Indonesia, for example, 9 million hectares have been allocated for development as industrial timber plantations, but only 2 million hectares have been replanted.[10]

Areas bereft of their original forest ecosystems and associated habitat have lost vegetation that stabilizes soil, cycles nutrients, and prevents erosion. These lands quickly lose utility and become a liability. Even when plantations are put in place, the functioning of a monoculture plantation is a far cry from that of an old-growth forest, where a number of species of differing ages each play a particular biological role, and ecosystem processes are thus bound to change.

A satellite-based survey of the world's forests by the U.N. Environment Programme, along with NASA and the U.S. Geological Survey, found that 80 percent of largely intact forests (those with a canopy closure of over 40 percent) are located in just 15 countries. A full 88 percent of the key closed forest areas are sparsely populated, making them hopeful targets for conservation. Short of calling for a moratorium of all logging, conservation in these 15 countries offers a reasonable starting point for forest preservation.[11]

Crucial to slowing the loss of the world's natural forests is finding alternative sources of energy for low-income countries, so that valuable wood is not burned. Innovations in reuse and recycling allow reclaimed timber and discarded paper to satisfy wood product demand.

Reduced consumption of virgin wood products is a key to saving the world's trees.

When wood products are used, governments can ensure that all domestic production and imports of wood products come from responsibly managed forests meeting rigorous environmental and social standards, like those of the Forest Stewardship Council (FSC). Worldwide, FSC-accredited bodies have certified some 24 million hectares of forests in 45 countries, numbers that are bound to increase as demand for certified wood rises and as noncertified sellers have difficulty competing.[12]

Water Scarcity Spreading

Lester R. Brown

Water scarcity may be the most underestimated resource issue facing the world today. As world water demand has more than tripled over the last half-century, signs of water scarcity have become commonplace. Some of the more widespread indicators are rivers running dry, wells going dry, and lakes disappearing.[1]

Among the rivers that run dry for part of the year are the Colorado in the United States, the Amu Darya in Central Asia, and the Yellow in China. China's Hai and Huai rivers have the same problem from time to time, and the flow of the Indus River—Pakistan's lifeline—is sometimes reduced to a trickle when it enters the Arabian Sea.[2]

The Colorado River, the largest in the southwestern United States, now rarely makes it to the sea. As the demand for water increased over the years, diversions from the river have risen to where they now routinely drain it dry.[3]

A similar situation exists in Asia, where the Amu Darya—one of the two rivers feeding the Aral Sea—now is dry for part of each year. With the sharp decline in the amount of water delivered to the Aral Sea by the Amu Darya, the sea has begun to shrink. There is a risk that the Aral could one day disappear entirely, existing only on old maps.[4]

China's Yellow River, the northernmost of its two major rivers, first ran dry for a few weeks in 1972. Since

1985, it has failed to make it to the Yellow Sea for part of almost every year. Sometimes the river does not even reach Shandong, the last province it flows through en route to the sea. As water tables have fallen, springs have dried up and some rivers have disappeared entirely. China's Fen River, the major watercourse in Shanxi Province, which once flowed through the capital of Taiyuan and merged with the Yellow, no longer exists.[5]

Another sign of water scarcity is disappearing lakes. In Central Africa, Lake Chad has shrunk by some 95 percent over the last four decades. Reduced rainfall, higher temperatures, and some diversion of water from the streams that feed Lake Chad for irrigation are contributing to its demise. In China, almost 1,000 lakes have disappeared in Hebei Province alone.[6]

Water tables are falling in several of the world's key farming regions, including under the North China Plain, which produces nearly one third of China's grain harvest; in the Punjab, which is India's breadbasket; and in the U.S. southern Great Plains, a leading grain-producing region.[7]

Water shortages now plague almost every country in North Africa and the Middle East. Algeria, Egypt, Iran, and Morocco are being forced into the world market for 40 percent or more of their grain supply. As population continues to expand in these water-short nations, dependence on imported grain is rising.[8]

Iran, one of the most populous countries in the Middle East, with 70 million people, is facing widespread water shortages. In the northeast, Chenaran Plain—a fertile agricultural region to the east of Mashad, one of Iran's largest and fastest-growing cities—is fast losing its water supply. Wells drawing from the water table below the plain are used for irrigation and to supply water to Mashad. The latest official estimate shows the water table falling by 8 meters in 2001 as the demand for water far outstrips the recharge rate of aquifers.[9]

Falling water tables in parts of eastern Iran have caused many wells to go dry. Some villages have been evacuated because there is no longer any accessible water. Iran is one of the first countries to face the prospect of water refugees—people displaced by the depletion of water supplies.[10]

In Yemen, a country of some 19 million people, water tables are falling everywhere by 2 meters or more a year. In the basin where the capital Sana'a is located, extraction exceeds recharge by a factor of five, dropping the water table by 6 meters (about 20 feet) a year. Recent wells drilled to a depth of 2 kilometers (1.3 miles) failed to find any water. In the absence of new supplies, the Yemeni capital will run out of water by the end of this decade.[11]

Another way of looking at water security is the amount of water available per person in a country. In 1995, 166 million people lived in 18 countries where the average supply of fresh water was less than 1,000 cubic meters a year—the amount deemed necessary to satisfy basic needs for food, drinking water, and hygiene. By 2050, water availability per person is projected to fall below the 1,000-cubic-meter benchmark in some 39 countries. By then, 1.7 billion people will in effect be suffering from hydrological poverty.[12]

At some point, the combination of aquifer depletion and the diversion of irrigation water to cities will likely begin to reduce the irrigated area worldwide. Data compiled by the U.N. Food and Agriculture Organization, based on official data submitted by governments, show irrigated area still expanding. For example, between 1998 and 1999, the last year for which global data are available, irrigated area grew from 271 to 274 million hectares. (See

Figure 2–5.) This reported 1-percent growth would be reassuring, but it appears to be overstated since governments are much better at gathering data on new irrigation projects than on irrigation reductions as water is diverted to cities or aquifers are depleted. It is quite possible that the historical growth in world irrigated area has come to a halt, and the area could even be declining.[13]

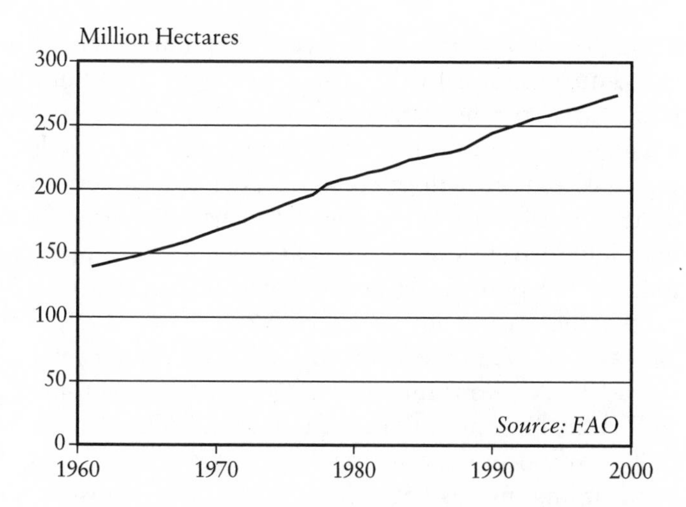

Figure 2–5. *World Irrigated Area, 1961–1999*

Carbon Emissions Climbing

Bernie Fischlowitz-Roberts

Though economic growth slowed throughout much of the world during 2001, world carbon emissions from burning fossil fuels continued their relentless upward trend, surpassing 6.5 billion tons. (See Figure 2–6.) As a result of the consistent growth of emissions, the atmospheric concentration of carbon dioxide (CO_2) has increased from the preindustrial level of 280 parts per million (ppm) to today's 370 ppm, a 32-percent increase. In the last 20 years, the atmospheric concentration of CO_2 has increased at the unprecedented rate of 1.5 ppm a year.[1]

In 1950, carbon emissions stood at 1.6 billion tons. By 1977, that had more than tripled, to 4.9 billion tons. In 2000, carbon emissions approached 6.5 billion tons, a quadrupling in just 50 years. Since the atmosphere's capacity to fix carbon is fairly constant, as the volume of emissions rises, the earth fixes a decreasing percentage of emissions. The increased atmospheric concentrations of CO_2 and other greenhouse gases (GHG) trap more of the earth's heat, causing temperatures to rise. These in turn are responsible for melting ice, rising sea levels, and a greater number of more destructive storms.[2]

Three fourths of the carbon emissions from human activities are due to the combustion of fossil fuels; the rest is caused by changes in land use, principally deforestation. Global energy consumption is projected to rise 60 percent over the next 20 years. Coal use is expected to

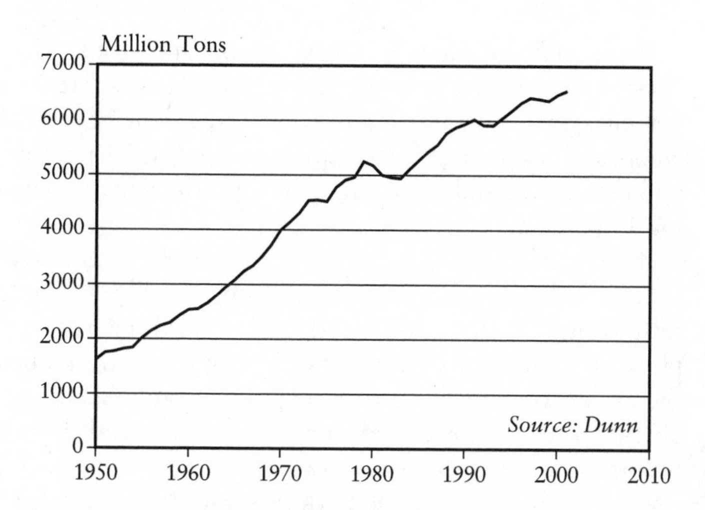

Figure 2–6. *World Carbon Emissions From Fossil Fuel Burning, 1950–2001*

increase by 45 percent, oil consumption by 58 percent, and natural gas by 93 percent, according to the U.S. Department of Energy. Since coal consumption has actually declined by 6 percent since its peak in 1996, however, there is reason to believe its use will either continue to drop or will increase less than projected. Yet even if coal usage remains steady over the next 20 years, the current level of emissions from all fossil fuels is simply too high. The increasing use of fossil fuels will only exacerbate changes in global climate.[3]

According to the Intergovernmental Panel on Climate Change (IPCC), atmospheric CO_2 concentrations by 2100 will be in the range of 650 to 970 ppm—more than double or triple preindustrial levels. As a result, the global average surface temperature will likely rise between 1.4 and 5.8 degrees Celsius between 1990 and 2100, an unprecedented rate of increase.[4]

Four major sectors produce carbon emissions. Electricity generation is responsible for the largest share—42 percent. Transportation generates 24 percent of global emissions. Industrial processes account for 20 percent, and residential and commercial uses produce the remaining 14 percent.[5]

Fortunately, changes can be made in each of these sectors to reduce carbon emissions using readily available technology. Shifting to wind, solar, and geothermal power for all electricity generation could greatly reduce the use of fossil fuels. Increased appliance and machinery efficiency could lower industrial and residential energy use. In the short term, shifts away from personal vehicles toward mass transit, along with increases in fuel efficiency, can reduce transportation emissions. And in the longer term, use of hydrogen-fueled cars and buses could cut emissions even further.

The United States is far and away the world's leading producer of carbon emissions, with 24 percent of the global total. China is responsible for 14 percent, and Russia accounts for 6 percent. Japan, whose economy is the second largest in the world, and India, whose population is second only to China, are each responsible for 5 percent of world emissions.[6]

Various policy measures have been put forward to address climate change and reduce concentrations of CO_2 and other greenhouse gases. The most prominent is the Kyoto Protocol, which commits industrial nations to reduce their emissions by at least 5 percent below 1990 levels by 2008–12. To enter into force, 55 countries representing 55 percent of emissions from industrial and former Eastern bloc nations must ratify the treaty. As of early June 2002, 74 countries responsible for 35.8 percent of global GHG emissions have ratified the protocol, including Japan and all nations of the European Union.

But with the United States and Australia refusing to ratify, the likelihood that it will enter into force is considerably diminished.[7]

In the United States, the Bush administration's "Clean Skies" proposal requires a decline in carbon emissions per unit of economic output (known as carbon intensity), but not overall carbon emissions. The flawed premise underlying the proposal is that economic growth cannot be achieved without significant carbon emission increases; thus "Clean Skies" will not fundamentally alter the U.S. emissions trajectory. The U.S. economy has consistently improved its carbon intensity, yet emissions have continued to increase. According to the American Council for an Energy-Efficient Economy, the carbon intensity of the U.S. economy was cut by 17 percent between 1990 and 2000, yet total emissions increased during that time by 14 percent due to a 39-percent increase in economic activity.[8]

The Kyoto Protocol, even if implemented, is only a first step. According to the IPCC, stabilizing atmospheric levels of CO_2 at 450 ppm would require fossil fuel emissions to drop below 1990 levels within a few decades, and eventually to decline to a small fraction of current levels.[9]

Regardless of the ultimate fate of the Kyoto Protocol, other policy initiatives show promise. Decreasing or eliminating government subsidies to fossil fuels, which total $300 billion annually worldwide, can move the energy economy away from heavy reliance on carbon-intensive fossil fuels. Decreasing taxes on income while instituting or increasing carbon taxes would constructively align economic and environmental goals. Increasing funding for further research and development of clean energy technologies can also help move the world from a carbon-based and toward a hydrogen-based energy system. Finally, stabilizing human population sooner rather than later will help reduce future emissions.[10]

Global Temperature Rising

Lester R. Brown

Last year, 2001, was the second warmest year since recordkeeping began in 1867. Following the all-time high of 1998, last year's near-record extends a strong trend of rising temperatures that began around 1980. The 15 warmest years since 1867 have all come since 1980. (See Figure 2–7.)[1]

This new year of temperature data provides further evidence that a trend of rising temperature is bringing to an end the period of relative climate stability that has prevailed since agriculture began some 11,000 years ago.

Monthly global temperature data compiled by NASA's Goddard Institute for Space Studies, in a series based on meteorological station estimates going back to 1867, show that September 2001 was the warmest September on record. November also set an all-time high. And six recent months—August and December 2001 and January, March, April, and May 2002—were each the second warmest respective months on record.[2]

The global average temperature for 2001 is calculated at 14.52 degrees Celsius (58.1 degrees Fahrenheit). The all-time high in 1998 was 14.69 degrees Celsius. Over the last century, the average global temperature climbed from 13.88 degrees Celsius in 1899–1901 to 14.44 degrees in 1999–2001, an increase of 0.56 degrees. But four fifths of this gain came in the century's last two decades.[3]

The rise of nearly 0.6 degrees Celsius during the last

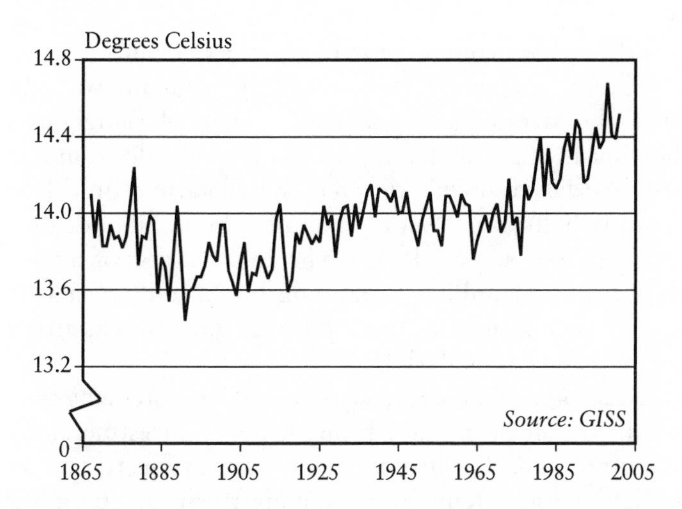

Figure 2–7. *Average Global Temperature, 1867–2001*

century is quite small compared with projections by the Intergovernmental Panel on Climate Change (IPCC) of the temperature rise for this century of 1.4–5.8 degrees Celsius (2.5–10.4 degrees Fahrenheit). Even the lower figure in that range would be more than double the increase of the last century. And the upper-end projection would be nearly 10 times as much.[4]

The contrast in sea level rise for the last century and that projected for this one is similarly worrying. During the last century, sea level rose an estimated 10–20 centimeters (4–8 inches). The IPCC projects that during this century sea level will rise 9–88 centimeters (4–35 inches).[5]

Rising temperature is not an irrelevant abstraction. It brings countless physical changes—from more intense heat waves, more severe droughts, and ice melting to more powerful storms, more destructive floods, and rising sea level. These changes in turn affect not only food security and the habitability of low-lying regions, but also the species composition of local ecosystems.

Climate change affects food security in many ways. In 2000, the World Bank published a map of Bangladesh showing that a 1-meter rise in sea level would inundate half of that country's riceland. Bangladesh would lose not only half its rice supply but also the livelihoods of a large share of its population. The combination of a population of 134 million expanding by 2.7 million a year and a shrinking cropland base is not a reassuring prospect for Bangladesh.[6]

Widespread changes in ecosystems are also being triggered. Recent years have brought heavy investments by governments and environmental organizations to protect particular ecosystems by converting them into parks or reserves. But if the rise in temperature cannot be checked, there is not an ecosystem on the earth that can be saved. Everything will change.

An additional year of temperature data reinforces the concerns expressed by the team of eminent scientists who produced the latest IPCC report, *Climate Change 2001*. They make clear what is now becoming obvious even to nonscientists: fossil fuel burning is changing the earth's climate.[7]

The bottom line is that altering the earth's climate is serious business—not something to be taken lightly. We can curb climate change by shifting from a carbon-based energy economy to one based on hydrogen. We have the technologies to do it. The economics are falling into place. Do we have the wisdom and the will to restructure the energy economy before climate change spirals out of control?

Ice Melting Everywhere

Lester R. Brown

Several new studies report that the earth's ice cover is melting faster than projected by the Intergovernmental Panel on Climate Change (IPCC) in its landmark report released in early 2001. (See Table 2–2.) Among other things, this means that the IPCC team, which did not have the ice melt data through the 1990s, will need to revise upward its projected rise in sea level for this century—currently estimated to range from 9 to 88 centimeters (4 to 35 inches).[1]

A study by two scientists from the University of Colorado's Institute of Arctic and Alpine Research found that melting of the large glaciers on the west coast of Alaska and in northern Canada is accelerating. Earlier data indicated that the melting of glaciers in these areas was raising sea level by 0.14 millimeters per year, but new data for the 1990s indicate that the more rapid melting is causing an increase of 0.32 millimeters a year, more than twice as fast.[2]

The Colorado study is reinforced by a U.S. Geological Survey (USGS) study that indicates glaciers are now shrinking in all 11 of Alaska's glaciated mountain ranges. An earlier USGS study reported that the number of glaciers in Glacier National Park in the United States had dwindled from 150 in 1850 to fewer than 50 today. It projected the remaining glaciers would disappear within 30 years.[3]

Another team of USGS scientists, using satellite data to

Table 2–2. *Selected Examples of Ice Melt Around the World*

Name	Location	Measured Loss
Arctic Sea Ice	Arctic Ocean	Over the last 35 years, ice has thinned from average of 3.1 meters to 1.8 meters. Could have ice-free summers before 2050.
Greenland Ice Sheet	Greenland	Has thinned by more than a meter a year on its southern and eastern edges since 1993.
Glacier National Park	Rocky Mtns., United States	Since 1850, the number of glaciers has dropped from 150 to fewer than 50. Remaining glaciers could disappear in 30 years.
Larsen B Ice Shelf	Antarctic Peninsula	Over the past five years has lost 5,700 square kilometers, 3,250 of which disintegrated in early 2002.
Ross Ice Shelf	Ross Sea	In March 2000, a piece of Ross Ice Shelf the size of Connecticut broke off, making one of the largest icebergs ever seen.

measure changes in the area covered by glaciers, describes an accelerated melting of glaciers in several mountainous regions, including the South American Andes, the Swiss Alps, and the French and Spanish Pyrenees.[4]

Glaciers are shrinking faster throughout the Andes. Professor Lonnie Thompson of Ohio State University reports that for the Qori Kalis glacier, on the west side of the Quelccaya ice cap in the Peruvian Andes, the annual shrinkage from 1998 to 2000 was three times that which occurred between 1995 and 1998. And that, in turn, was nearly double the annual rate of retreat from 1993 to

Table 2–2 *continued*

Dokriani Bamak Glacier	Himalayas, India	Retreated by 20 meters in 1998, compared with 16.5 meters over the previous five years.
Tien Shan Mountains	Central Asia	In the past 40 years, 22 percent of glacial volume has vanished.
Caucasus Mountains	Russia	Glacial volume has declined by 50 percent in the past century.
Alps	Western Europe	Glacial volume has shrunk by more than 50 percent since 1850.
Kilimanjaro	Tanzania	Ice cap shrunk by 33 percent from 1989 to 2000. Could disappear by 2015.
Quelccaya Ice Cap	Andes, Peru	Rate of retreat increased to 30 meters a year in the 1990s, up from only 3 meters a year; will likely disappear before 2020.

Source: See endnote 1.

1995. Thompson projects that the large Quelccaya ice cap will disappear entirely between 2010 and 2020.[5]

The vast snow/ice mass in the Himalayas, which ranks third after Antarctica and Greenland in the amount of fresh water stored, is also retreating. Although data are not widely available for the Himalayan glaciers, those that have been studied indicate an accelerating retreat. For example, data for the 1990s show that the Dokriani Bamak Glacier in the Indian Himalayas moved back by 20 meters in 1998 alone, more than during the preceding five years.[6]

Thompson has also studied Kilimanjaro, observing that between 1989 and 2000, this famous mountain in

Tanzania lost 33 percent of its ice field. He projects that the ice could disappear entirely within the next 15 years.[7]

Both the North and the South Poles are showing the effects of climate change too. The South Pole is covered by a continent the size of the United States. The Antarctic ice sheet, which is 2.5 kilometers (1.5 miles) thick in some places, contains over 70 percent of the world's fresh water and 90 percent of the earth's ice.[8]

While this vast ice sheet is relatively stable, the ice shelves—the portions of the ice sheet that extend into the surrounding seas—are fast disappearing. Over the past five years, the Larsen B ice shelf on the Antarctic Peninsula has lost more than 5,700 square kilometers of ice, half of which disappeared in the early months of 2002. Delaware-sized icebergs that have broken off are a threat to ships in the area.[9]

While the South Pole is covered by a huge continent, the North Pole is covered by the Arctic Ocean. Arctic sea ice is melting fast. Over the last 35 years, the ice has thinned 42 percent—from an average of 3.1 meters to 1.8 meters. It has also shrunk by 6 percent since 1978. Together, thinning and shrinking have reduced the mass of sea ice by half. A team of Norwegian scientists projects that the Arctic Sea could be entirely ice-free during the summer by mid-century, if not before.[10]

If this melting materializes as projected, the early explorers' dream of a northwest passage—a shortcut from Europe to Asia—could be realized. Unfortunately, what was a dream for them could be a nightmare for us.

If the Arctic Ocean becomes ice-free in the summer, it would not affect sea level because the ice is already in the water, but it would alter the regional heat balance. When sunlight strikes ice and snow, most of it is reflected back into space, but if it strikes land or open water, then much of the energy in the light is absorbed, leading to higher temperatures. This is what computer modelers refer to as a positive feedback loop, a situation where a trend creates self-reinforcing conditions.

Richard Kerr, writing in *Science*, notes that summer "would convert the Arctic Ocean from a brilliantly white reflector sending 80 percent of solar energy back into space into a heat collector absorbing 80 percent of [incoming sunlight]." The discovery of open water at the North Pole by an ice breaker cruise ship in August 2000 provides further evidence that the melting process may now be feeding on itself.[11]

This prospect of much warmer summers in the Arctic is of concern because Greenland, which has the world's second largest ice sheet, is largely within the Arctic Circle. In a *Science* article in 2000, a team of U.S. scientists from NASA reported that the vast Greenland ice sheet is starting to melt.[12]

The team also reports that the melting there appears to be accelerating because the ice sheet on its southern and eastern edges has thinned by more than a meter a year since 1993. If all the ice on Greenland were to melt, it would raise sea level by 7 meters (23 feet), but even under a high temperature rise scenario, it could take many centuries for it to melt completely.[13]

The accelerated melting of ice, particularly during the last decade or so, is consistent with the accelerating rise in temperature that has occurred since 1980. With the IPCC projecting global average temperature to rise by 1.4–5.8 degrees Celsius (2.5–10.4 degrees Fahrenheit) during this century, the melting of ice will likely continue to gain momentum.[14]

Our generation is the first to have the capacity to alter the earth's climate. We are also, therefore, the first to wrestle with the ethical question of whether the capacity to change the planet's climate gives us the right to do so.

Wind Electric Generation Soaring

Lester R. Brown

World wind electric generating capacity climbed from 17,500 megawatts (MW) in 2000 to 24,000 MW in 2001—a dramatic one-year gain of 6,500 MW or 37 percent. As generating costs continue to fall and as public concern about climate change escalates, the world is fast turning to wind for its electricity.[1]

Since 1995, world wind generating capacity has increased an astounding fivefold. In stark contrast, the use of coal—the principal alternative for generating electricity—peaked in 1996 and has declined by 6 percent since then.[2]

One megawatt of wind generating capacity typically will satisfy the electricity needs of 350 households in an industrial society, or roughly 1,000 people. Thus, the 24,000 MW of generating capacity now in place is sufficient to meet the residential electricity needs of some 24 million people—equal to the combined populations of Denmark, Finland, Norway, and Sweden.

In wind electric generating capacity, Germany leads the world with 8,750 MW, more than a third of the total. The United States, which launched the modern wind power industry in California in the early 1980s, follows with 4,250 MW. Spain is in third place, with 3,300 MW. Denmark, which is fourth with 2,400 MW, now gets more than 15 percent of its electricity from wind. Almost two thirds of the capacity added in 2001 was concentrated in the top three countries: Germany added 2,600 MW; the United States, 1,700; and Spain, 930. For the United States, this translates into a growth in generating capacity of some 67 percent in 2001.[3]

Despite this spectacular growth, development of the earth's wind resources has barely begun. In densely populated Europe, there is enough easily accessible offshore wind energy to meet all of the region's electricity needs. In the United States, the wind-rich states in the Great Plains have enough harnessable wind energy to meet the country's electricity needs. And China can easily double its current electricity generation from wind alone.[4]

The cost of wind-generated electricity at prime wind sites has fallen dramatically in the United States over the last 15 years—from 35¢ per kilowatt-hour in the mid-1980s to 4¢ per kilowatt-hour in 2001. (See Figure 2–8.) A few long-term supply contracts have even been signed recently for 3¢ per kilowatt-hour. With the U.S. adoption of a wind production tax credit (PTC) in 1993 to offset established subsidies for oil, coal, and nuclear power, growth surged. New wind farms came online in Colorado, Iowa, Kansas, Minnesota, New York, Oregon, Pennsylvania, Texas, Washington, and Wyoming. In March 2002, the PTC was extended until the end of 2003, setting the stage for continuing rapid growth.[5]

Low-cost electricity from wind brings the option of electrolyzing water to produce hydrogen, which can easily be stored and used to fuel gas-fired turbines in backup power plants when wind power ebbs. Over time, hydrogen produced with wind-generated electricity is the leading candidate to replace natural gas in gas-fired power plants as gas reserves are depleted.

Hydrogen is also the ideal fuel for the fuel-cell engines that every major automobile manufacturer is now working on. Honda and DaimlerChrysler both plan to have

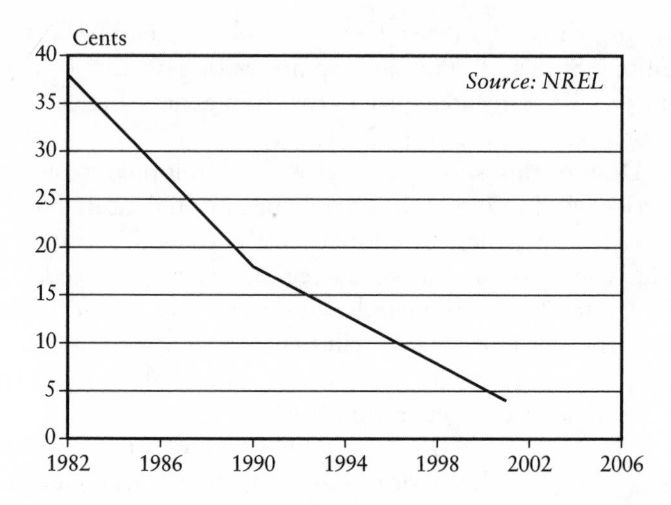

Figure 2–8. *Average Cost Per Kilowatt-Hour of Wind-powered Electricity in the United States, 1982–2001*

fuel-cell-powered vehicles on the market in 2003.[6]

Wind power offers long-term price stability and energy independence. Not only are costs low and falling, but with wind-generated electricity there are no abrupt price hikes, as there are with natural gas. There is no OPEC for wind, because wind is widely dispersed. An inexhaustible source of energy, wind offers more energy than society can use, and it does not disrupt climate.

Investment in wind turbine manufacture and wind development has been highly profitable. While high-tech firms as a group suffered a disastrous fall in sales, earnings, and stock value in 2001, sales in the wind industry soared. At Danish-based Nordex, for example, one of the world's largest turbine manufacturers, turnover during the first half of fiscal year 2001/2002 was up 47 percent.[7]

Even more impressive than the recent growth in generating capacity are the plans for future growth. The

European Wind Energy Association has recently revised its 2010 wind capacity projections for Europe from 40,000 megawatts to 60,000 megawatts.[8]

France, which for years had ignored wind power, announced in December 2000 that it would develop 5,000 megawatts of wind-generating capacity during this decade. A few weeks later, Argentina announced it was planning to develop 3,000 megawatts of wind-generating capacity in Patagonia. In April 2001, the United Kingdom sold offshore lease rights for an estimated 1,500 megawatts of wind-generating capacity to several different bidders, including Shell Oil. And in early 2002, China announced plans to develop up to 1,200 megawatts of wind capacity by 2005.[9]

In the United States, wind generating capacity is growing by leaps and bounds. The 261-megawatt Stateline Wind Project on the border between Oregon and Washington will be expanded to 300 megawatts later this year, making it the world's largest wind farm. Texas added some 900 megawatts in several projects during 2001, including a 278-MW wind farm at King Mountain in west Texas, currently the world's largest. In South Dakota, Jim Dehlsen, a pioneer in developing California's wind energy, has secured the wind rights to 90,000 hectares (222,000 acres) of farm and ranchland in the east central part of the state. He plans to develop a huge 3,000-megawatt wind farm and to transmit the electricity across Iowa, supplying Illinois and other states in the industrial Midwest.[10]

In Europe, offshore projects are now springing up off the coasts of Belgium, Denmark, France, Germany, Ireland, the Netherlands, Scotland, Sweden, and the United Kingdom.

The German Wind Energy Institute projects installation of 2,900 MW in 2002, and 2,400 MW in 2003. If these installations materialize as projected, total installed capacity in the country will easily surpass the German government's 2010 goal of 12,500 MW by the end of 2003.[11]

Projecting future growth in such a dynamic industry is complicated, but once a country has developed 100 megawatts of wind-generating capacity, it tends to move quickly to develop its wind resources. The United States crossed this threshold in 1983. In Denmark, this occurred in 1987. In Germany, it was 1991, followed by India in 1994 and Spain in 1995.[12]

By the end of 1999, Canada, China, Italy, the Netherlands, Sweden, and the United Kingdom had also all crossed this threshold. During 2000, Greece, Ireland, and Portugal joined the list. And in 2001, it was France and Japan. So as of early 2002, some 16 countries—home to half the world's people—have entered the fast-growth phase in wind power development.[13]

Wind energy in the form of electricity and hydrogen can satisfy all the various energy needs of a modern economy, and it promises to become the foundation of the new energy economy. We can now see the shape of this new economy emerging as wind turbines replace coal mines, hydrogen generators replace oil refineries, and fuel-cell engines replace internal combustion engines.

Bicycle Production Breaks 100 Million

Janet Larsen

Over 100 million bicycles were manufactured in 2000, the most since the all-time high of 106 million in 1995. (See Figure 2–9.) This production level is double that of 25 years ago.[1]

China manufactured a record 52 million bicycles in 2000—over half the world total. Nearly two thirds of these were exported, with 17 million going to the United States. The United States itself produced just over 1 million bikes, down sharply from the 1995 output of nearly 9 million. With over 43 million cyclists, the United States is the world's largest bicycle export market, with imports meeting 97 percent of demand.[2]

The European Union, led by Germany, produced some 12 million bicycles in 2000. Italy closely trails German production of 3.2 million bicycles, although cycle sales in Germany reached 5.3 million in 2000, compared with 1.6 million units in Italy.[3]

India produced more than 11 million bicycles. Most of these are ridden domestically or shipped to Africa. Africa is a potentially large bicycle market, but recently sales have declined in many countries despite the continued need for low-cost, non-motorized transportation. One reason for this trend is a shortage of moderately priced, modern bikes and bike parts.[4]

This shortage is seen in Senegal, which levies prohibitive tariffs on imported cycles to protect a small

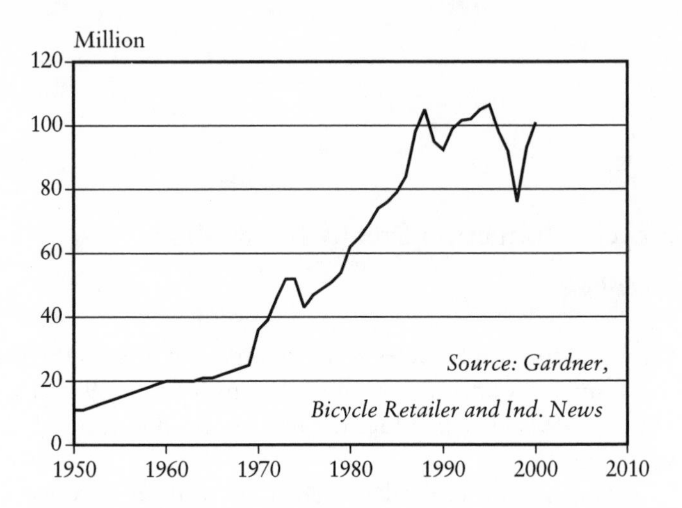

Figure 2–9. *World Bicycle Production, 1950–2000*

domestic manufacturer that sells only 2,000 bikes annually. Until 1989, Ghana imposed similar tariffs and taxes on imports, but after their removal, bike sales soared.[5]

To meet Africa's high demand for modern and sturdy bicycles, the Institute for Transportation and Development Policy, an organization that promotes environmentally sustainable and equitable transportation policies worldwide, and Afribike, a nonprofit South African company, designed the Africa Bike. This is an alternative to the traditional Black Roadster, which now sells poorly because it does not meet performance standards and because many associate it with rural, elderly, and poor people. Both models retail at about $60. Afribike alone has provided over 10,000 South Africans with low-cost transportation since 1998 and plans to expand its programs to Senegal, Guinea, and Ghana.[6]

Bicycle ownership greatly enhances personal mobility, contributing to substantial increases in income. Giving women in rural areas credit to buy bicycles allows for increased access to education and facilitates the transport of produce to market. Thus rising bicycle sales have been correlated with higher farm output. In Ghana, bikes have helped HIV/AIDS outreach workers reach 50 percent more beneficiaries.[7]

In urban areas, bicycles can substitute for automobiles, reducing traffic congestion and lowering air pollution and noise. Bicycles take up one thirtieth the road space used by cars traveling at a moderate pace. Biking also offers exercise at a time when more people are overweight or obese than ever before, even in developing countries. For those who need help traveling long distances or in hilly terrain, increasingly popular electric bicycles that run on batteries often fit the bill. By 2003, bicycles powered by fuel cells will hit the market.[8]

A number of cities, particularly in industrial countries, are promoting the bicycle as a sustainable form of transportation by developing cycleways and offering incentives for using bicycles for commuting. In Copenhagen, one third of the population commutes to work by bicycle. By 2005, Copenhagen's innovative City Bike program will provide 3,000 bicycles for free use within the city. The city's total cycle fleet is expected to grow, as city planners intend to increase already high car parking fees by 3 percent annually for 15 years, impose high fuel taxes and vehicle registration costs, and concentrate future development around rail lines.[9]

Stockholm, one of the world's wealthiest cities, has seen car use decline in recent decades. There, urban development is concentrated around city centers, allowing for greater public transportation efficiency. Rail and buses are linked with pedestrian and bicycle-oriented routes. In all of Sweden's urban areas, 1 out of every 10 trips is taken by bicycle, about the same number by public transit, and almost 40 percent on foot. Just 36 percent of trips are taken by car, a low for Europe. In the Netherlands, bicycles are used for 27 percent of all trips.[10]

Yet with the world automobile fleet climbing to over 530 million, bicycles are losing out to a growing collection of motorized vehicles in some parts of the world. In Beijing 10 years ago, 60 percent of all trips were made on bicycle. Now that incomes have risen, residents have begun to favor the car, which is viewed as a symbol of progress, and bike trips have fallen to 40 percent. In Shanghai, where many major streets have recently been closed to bicycles during rush hour, the share of trips made by bike has dropped to 20 percent. The Shanghai government reportedly has plans to ban bicycles altogether from the city center by 2010.[11]

In the United States and Canada, where development is much less concentrated, 84 and 74 percent of trips are made by car respectively. In both countries, only about 10 percent of trips are pedestrian, and just 1 percent is by bicycle. Many residents use bicycles for recreation, not for transit.[12]

Cities at risk of being overrun by polluting, land-hungry automobiles could benefit by ensuring that bicycles receive consideration in transportation planning and urban development schemes. Tax incentives can encourage development in areas close to mass transit, and trains and buses can be equipped to carry bicycles. Making streets and pathways safer and accessible to cyclists will encourage more people to pedal to work and to use bikes for recreation.

Annual world bicycle production has grown to more than double automobile production since the mid-twentieth century, when the two nearly coincided. The bicycle is an affordable, space-efficient, low-maintenance method of personal transportation, and its usefulness promises future growth in the industry.[13]

Solar Cell Sales Booming

Bernie Fischlowitz-Roberts

In 2001, world solar cell production soared to 395 megawatts (MW), up 37 percent over 2000. (See Figure 2–10.) This annual growth in output, now comparable in size to a new power plant, is set to take off in the years ahead as production costs fall. Cumulative solar cell or photovoltaic (PV) capacity now exceeds 1,840 MW.[1]

The top five producers in 2001 were Sharp, BP Solar, Kyocera, Siemens Solar, and AstroPower, accounting for 64 percent of global output. Japanese manufacturers, with 43 percent of the world total, benefited from government policies to encourage solar cell use. The 70,000 Roofs Program, which initially provided a 50-percent cash subsidy for grid-connected residential systems, has been the primary driver of Japan's PV market expansion. The subsidy declined to 35 percent in 2000 as production increased and solar cell prices dropped. In addition to residential subsidies, government spending of $271 million in fiscal year 2001—on research and development, demonstration programs, and market incentives—was key to the growth.[2]

In contrast to Japan, the U.S. government spent only $60 million on solar programs in 2000. The U.S. share of the global market—24 percent—was surpassed in 2001 by the European Union (EU), which now accounts for 25 percent. Government commitments to renewable energy are more robust in the EU than in the United States. In

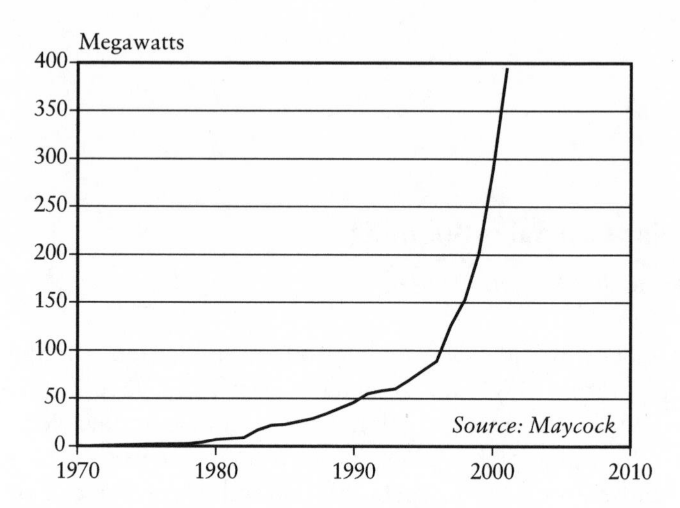

Figure 2–10. *World Photovoltaic Shipments, 1971–2001*

Germany, the Renewable Energy Act of 2000 offers citizens preferable loan terms for purchasing solar systems, and gives them a guaranteed price when feeding excess energy back into the power grid (known as net metering). As a result of such support, the German PV industry—the most advanced in Europe—is projected to grow from its current installed capacity of 113 MW in 2001 to 438 MW by 2004.[3]

Due to government policies in Japan, grid-connected residential installations totaling 100 MW dominated sales in 2001. Germany's grid-connected systems accounted for around 75 MW. The 32 MW installed in the United States were divided between grid-connected systems and those in remote areas not linked to a power grid. All of India's 18 MW were for such off-grid installations. The 120–130 MW installed in some 50–60 developing nations were also for off-grid projects.[4]

Both Japan and the United States were net exporters

of solar cells. Almost two thirds of U.S. output was exported, while Japan exported 42 percent of its total.[5]

The cost of electricity from solar cells remains higher than from wind or coal-fired power plants for grid-connected customers, but it is falling fast due to economies of scale as rising demand drives industry expansion. Solar cells currently cost around $3.50 per watt for crystalline cells, and $2 per watt for thin-film wafers, which are less efficient but can be integrated into building materials. Industry analysts note that between 1976 and 2000, each doubling of cumulative production resulted in a price drop of 20 percent. Some maintain that prices may fall even more dramatically in the future.[6]

The European Photovoltaic Industry Association suggests that grid-connected rooftop solar systems could account for 16 percent of electricity consumption in the 30 members of the Organisation for Economic Co-operation and Development by 2010. If costs of rooftop PV systems fall to $3 per watt by the middle of this decade, as projections suggest, the market for residential rooftop solar systems will expand. In areas where home mortgages finance PV systems and where net metering laws exist, demand could reach 40 gigawatts, or 100 times global production in 2001.[7]

More than a million homes worldwide, mainly in villages in developing countries, now get their electricity from solar cells. For the 1.5–2 billion people whose homes are not connected to an electrical grid, solar cells are typically the cheapest source of electricity. In remote areas, delivering small amounts of electricity through a large grid is cost-prohibitive, so people not close to a grid will likely obtain electricity from solar cells. If microcredit financing is arranged, the monthly payment for photovoltaic systems is often comparable to what a family would spend on candles or on kerosene for lamps.

After the loan is paid off, typically in two to four years, the family obtains free electricity for the remainder of the system's life.[8]

Photovoltaic systems furnish high-quality electric lighting, which can improve educational opportunities, provide access to information, and help families be more productive after sunset. A shift to solar energy also brings health benefits. Solar electricity allows for the refrigeration of vaccines and other essentials, playing a part in improving public health. For many rural residents in remote areas, a shift to solar electricity improves indoor air quality. PV systems benefit outdoor air quality as well. The replacement of a kerosene lamp with a 40-watt solar module eliminates up to 106 kilograms of carbon emissions a year.[9]

In addition to promising applications in the developing world, solar also benefits industrial nations. Even in the United Kingdom, a cloudy country, putting modern PV technology on all suitable roofs would generate more electricity than the nation consumes in a year. This would eliminate all greenhouse gas emissions from nationwide electricity generation, removing almost 200 million tons of carbon dioxide annually from the atmosphere.[10]

Recent research on zero-energy homes, where solar panels are integrated into the design and construction of extremely energy-efficient new houses, presents a promising opportunity for increased use of solar cells. Julius Poston, a progressive builder in the southeastern United States, builds homes that use half the energy of typical ones. His company, Certified Living, has constructed two prototype zero-energy homes with integrated solar panels. If eventually adopted on a wide scale, this groundbreaking concept could eliminate the pollution associated with fossil fuel–generated electricity for households.[11]

Continued strong growth suggests that the solar cell market will play a prominent role in providing renewable, non-polluting sources of energy in both developing and industrial countries. A number of policy measures can help ensure the future growth of solar power. Removing distorting subsidies of fossil fuels would allow solar cells to compete in a more equitable marketplace. Expanding net metering laws to other countries and the parts of the United States that currently do not have them will make owning solar home systems more economical by requiring utilities to buy electricity back from homeowners. Finally, revolving loan funds and other providers of microcredit are essential to the rapid spread of solar cell technologies in developing nations.

Solar cell manufacturers are beginning to sense the enormous growth in the market that lies ahead. Japan-based Sharp Corporation, already the world's leading producer of solar cells, plans to double its capacity in 2002, going from 94 to 200 megawatts. For the industry as a whole, output is expected to increase at 40–50 percent annually over the next few years, bringing the solar age ever closer.[12]

3

Eco-Economy Updates

Some caveats are in order in introducing this section. The following 20 Eco-Economy Updates are printed as they appeared when first released. They have not been revised or updated.

In at least one instance, there has been a dramatic change. "The Rise and Fall of the Global Climate Coalition," was written in July 2000. It describes how leading corporations were abandoning the Global Climate Coalition (GCC) because of philosophical differences and the negative public relations effect of being associated with the group. GCC's massive advertising campaign to prevent the United States from endorsing any meaningful agreement to reduce global carbon emissions, specifically the Kyoto Protocol, was out of sync with the views of many of its leading corporate members. The update describes how Dupont, British Petroleum, then Royal Dutch Shell, and then Ford, DaimlerChrysler, Texaco, General Motors, and other leading corporations were leaving the Coalition.

What we did not anticipate at the time was that in January 2002 the Coalition would close its doors. It no longer exists. We still include the update in the collection because it reflects the evolution of thinking within the corporate community on global climate issues. At some point, there was simply no longer a place for the Global Climate Coalition.

There is also repetition, both among the updates in this section and between these updates and information in Parts 1 and 2 of *The Reader*. We apologize for this, but again we did not feel it appropriate to rewrite the updates, partly because to exclude some of the repetitive information would have weakened their structure.

One of the key points about wind energy that is repeated, for example, is its enormous potential. To illustrate this we point out that 3 of the 50 states in the United States—Kansas, North Dakota, and Texas—have enough harnessable wind energy to satisfy national electricity needs. If you did not know this when you started to read this book, you will when you finish!

Backup graphs and tables for all the updates, as well as links to additional information, are included on our Web site. At the end of each update we have listed its particular Web address.

With these caveats in mind, we are pleased to present Part 3 of *The Earth Policy Reader*. If you wish to receive future Eco-Economy Updates free of charge you can do so by visiting us at <www.earth-policy.org> and subscribing to our listserv. You can also e-mail your request to <epi@earth-policy.org>.

Energy and Climate

June 2000

U.S. Farmers Double Cropping Corn and Wind Energy

Lester R. Brown

Farmers and ranchers in the United States are discovering that they own not only land but also the wind rights that go with that land. A farmer in Iowa who leases a quarter acre of cropland to the local utility as a site for a wind turbine can typically earn $2,000 a year in royalties from the electricity produced. In a good year, that same plot can produce $100 worth of corn. Wind turbines strung across the farm at appropriate intervals can provide a welcome boost to farm income, yielding a year-round cash flow.[1]

Harnessing the wind has become increasingly profitable. The American Wind Energy Association reports that the cost per kilowatt-hour of wind-generated electricity has fallen from 38¢ in the early 1980s to 3–6¢ today, depending primarily on wind speed at the site. Already competitive with other sources, the cost of wind-generated electricity is expected to continue to decline. These falling costs, facilitated by advances in wind turbine design, help explain why wind power is expanding rapidly beyond its original stronghold in California.[2]

As wind farms have come online in farming and ranching states such as Iowa, Minnesota, Texas, and Wyoming, wind electric generation has soared, pushing U.S. wind generating capacity from 1,928 megawatts in 1998 to 2,490 megawatts in 1999—a gain of 29 percent. Contrary to public perceptions, the potential of wind power is enormous. A U.S. Department of Energy wind resource inventory found that three states—Kansas, North Dakota, and Texas—have enough harnessable wind energy to meet electricity needs for the whole country.[3]

At a time when farmers are struggling with low grain prices, some are now finding salvation in this new "crop," enabling them to stay on the land. It is like striking oil, except that the wind is never depleted.

In the Great Plains, where an acre of rangeland produces only $20 worth of beef a year or an acre in wheat may yield $120 worth of grain, the attraction of wind power is obvious. For ranchers with prime wind sites, income from wind could easily exceed that from cattle sales.[4]

One of the attractions of wind energy is that the turbines scattered across a farm or ranch do not interfere with the use of the land for farming or cattle grazing. Farmers can literally have their cake and eat it too.

Another attraction is that much of the income generated stays in the local community, whereas if electricity comes from an oil-fired power plant, the money spent for electricity may end up in the Middle East. With a single large wind turbine generating $100,000 or more worth of electricity per year, harnessing local wind energy can revitalize rural communities.[5]

And it is not only the wind farms themselves that provide income, jobs, and tax revenue. The first utility-scale wind turbine manufacturing facility to be built outside of California has recently started operation in Champaign, Illinois, in the heart of the Corn Belt.[6]

Agricultural land values may soon reflect this new source of income. The wind meteorologist who identifies the best sites for turbines is playing a role in the emerging new energy economy comparable to that of the petroleum geologist in the old energy economy. The mere sight of a wind meteorologist installing wind-measuring instruments in a community could raise land prices.

Satisfying the local demand for electricity from wind is not the end of the story. Cheap electricity produced from wind can be used to electrolyze water, producing hydrogen, now widely viewed as the fuel of the future. With automobiles powered by fuel-cell engines expected on the market within a few years and with hydrogen as the fuel of choice for these new engines, a huge new market is opening up. Royal Dutch Shell, a leader in this area, is already starting to open hydrogen stations in Europe. William Ford, CEO of the Ford Motor Company, has said he expects to preside over the demise of the internal combustion engine.[7]

Farms and ranches may one day supply the hydrogen that will power the nation's motor vehicle fleet, giving the United States the energy source needed to declare its independence from Middle Eastern oil.

Concerned about burning fuels that destabilize climate, government at all levels is encouraging the development of climate-benign renewable energy sources. In some states, utility commissions are requiring utilities to offer their customers a "green power" option. Although this usually means a slightly higher monthly electricity bill, many consumers worried about climate change are choosing green power. In Colorado, offering a wind power option to both residential and business electricity users has already led to the installation of 20 megawatts of wind generating capacity—an amount expected to double soon.[8]

Many state governments are taking the initiative. Minnesota is requiring its largest utility to install 425 megawatts of wind-generating capacity by 2002. In Texas, the legislature has set a goal of 2,000 megawatts of generating capacity from renewable sources by 2009, with most of it expected to come from the Lone Star state's abundant wind power.[9]

At the national level, U.S. Secretary of Energy Bill Richardson is requiring that 7.5 percent of the electricity used in his Department come from renewable sources (excluding hydro) by 2010.[10]

A formidable new alliance is emerging in support of wind energy. In addition to environmentalists, farmers and those consumers who favor green power are now supporting the development of the nation's wind wealth. So, too, are political leaders in the farming and ranching states of the Midwest and the Great Plains, many of whom sponsored legislation in Washington to extend the wind energy production tax credit, which encourages investment in wind power. Aside from the economic benefits of wind power, political interest is being spurred by a steady diet of news stories about the possible effects of global warming, including record heat waves and droughts, melting glaciers, and rising sea level.

Rapid growth in wind energy is not limited to the United States. Worldwide, wind electric generation in 1999 expanded by a staggering 39 percent. Wind already supplies 10 percent of Denmark's electricity. In Germany's northernmost state of Schleswig-Holstein, it supplies some 14 percent of all electricity. Spain's northern industrial province of Navarra gets 23 percent of its electricity from wind, up from zero just four years ago. In

China, which recently brought its first wind farm online in Inner Mongolia, wind analysts estimate that the country's wind potential is sufficient to double national electricity generation.[11]

In Denmark, Germany, and the Netherlands, individual farmers, or organized groups of farmers, are investing in the turbines themselves and selling the electricity to the local utilities, thus boosting the farmers' share of income from wind power.[12]

The world is beginning to recognize wind for what it is—an inexhaustible energy source that can supply both electricity and fuel. In the United States, farmers are learning that two crops are better than one, political leaders are realizing that harnessing the wind can contribute to both energy security and climate stability, and consumers are finding out that they can help stabilize climate. This is a winning combination—one that will help make wind energy a cornerstone of the new energy economy.

For additional information, see <www.earth-policy.org/Alerts/Alert3.htm>.

July 2000

The Rise and Fall of the Global Climate Coalition*

Lester R. Brown

In August 1997, a few months before the Kyoto conference on climate change, the Global Climate Coalition (GCC) helped launch a massive advertising campaign designed to prevent the United States from endorsing any meaningful agreement to reduce global carbon emissions. This group, including in its ranks some of the world's most powerful corporations and trade associations involved with fossil fuels, concentrated its efforts on a series of television ads that attempted to confuse and frighten Americans.[1]

Among other things, the ads indicated that "Americans will pay the price... 50¢ more for every gallon of gasoline," even though there was no proposal for such a tax. The campaign was successful. The so-called Carbon Club had effectively undermined public support of U.S. efforts to lead the international effort to stabilize climate.[2]

While the public image of the GCC at the time was that of a unified group, there was already dissent within the ranks. John Browne, Chairman of British Petroleum, in a speech at Stanford University on May 19, 1997, announced that "the time to consider the policy dimensions of climate change is not when the link between greenhouse gases and climate change is conclusively

* The GCC officially closed its doors on 28 January 2002 after most of its leading corporate members had abandoned it.

proven, but when the possibility cannot be discounted and is taken seriously by the society of which we are part. We in BP have reached that point."[3]

Browne's talk shocked other oil companies and pleasantly surprised the environmental community. BP withdrew from the Global Climate Coalition. Dupont had already left. The following year, Royal Dutch Shell announced that it, too, was leaving. Its corporate goals, like those of BP and Dupont, no longer meshed with those of the GCC. Like BP, it no longer viewed itself as an oil company, but as an energy company.[4]

In 1999, Ford withdrew from the GCC. Its young Chairman, William C. Ford, Jr., the great-grandson of Henry Ford, went on record saying, "I expect to preside over the demise of the internal combustion engine." The company was already working on a fuel-cell engine, one where the fuel of choice was hydrogen—not gasoline.[5]

Ford's decision to withdraw was yet another sign of the changes occurring in major industries involved directly and indirectly with fossil fuels. A company spokesman noted, "Over the course of time, membership in the Global Climate Coalition has become something of an impediment for Ford Motor to credibly achieving our environmental objectives."[6]

In rapid succession in the early months of 2000, DaimlerChrysler, Texaco, and General Motors announced that they too were leaving the Coalition. With the departure of GM, the world's largest automobile company, the die was cast. A spokesman for the Sierra Club quipped, "The only question left is whether the last one out of GCC will turn off the light."[7]

The image created by this accelerating exodus of firms from the GCC was that of rats abandoning a sinking ship. It reflected the conflict emerging within GCC ranks between firms that were clinging to the past and those that were planning for the future.

Some of the exiting companies, such as BP Amoco, Shell, and Dupont, joined a progressive new group, the Business Environmental Leadership Council, now an organization of some 21 corporations. This new outfit, founded by the Pew Center on Global Climate Change, says, "We accept the views of most scientists that enough is known about the science and environmental impacts of climate change for us to take actions to address its consequences."[8]

Other leading companies that have joined the Council are Toyota, Enron, and Boeing. Membership requires individual companies to have their own programs for reducing carbon emissions. BP Amoco, for example, plans to bring its carbon emissions to 10 percent below its 1990 level by 2010, exceeding the Kyoto goal of roughly 5 percent for industrial countries.[9]

Dupont has one of the most ambitious goals of any company, going far beyond that of Kyoto. It has already cut its 1990 greenhouse gas emissions by 45 percent and plans to reduce them by a total of 65 percent by 2010, rendering hollow the claim that lowering carbon emissions to meet the Kyoto goal is not possible.[10]

On the supply side, BP Amoco and Shell are investing heavily in new sources of energy. BP Amoco is now a leading manufacturer of solar cells. Shell, already a major player in both wind and solar cells, is also investing heavily in hydrogen and will likely open the world's first chain of hydrogen stations in Iceland.[11]

To date, the net effect of the various public and private initiatives worldwide has been to check the growth in global carbon emissions. Since 1996, global carbon emissions have leveled off. The burning of coal, the most carbon-intensive fuel, dropped 5 percent in 1999. The next step is to reduce carbon emissions across the board.[12]

Abandonment of the Global Climate Coalition by leading companies is partly in response to the mounting evidence that the world is indeed getting warmer. The 15 warmest years in the last century have occurred since 1980. Ice is melting on every continent. The snow/ice pack in the Rockies, the Andes, the Alps, and the Himalayas is shrinking. The volume of the ice cap covering the Arctic Ocean has shrunk by more than 40 percent over the last 35 years. To deny that the earth is getting warmer in the face of such compelling evidence is to risk a loss of credibility, something that corporations cannot readily afford.[13]

The high price paid by the tobacco industry's continuing denial of a link between smoking and health is all too familiar. This loss of credibility led to a major shift in public opinion, one that is now affecting court proceedings and the decisions of juries considering the claims of plaintiffs against the tobacco industry. And it figured prominently in the agreement by the industry to pay state governments $251 billion to compensate them for the Medicare costs of treating smoking-related illnesses.[14]

In a thinly veiled effort to conceal the real issue—the loss of so many key corporate members—the GCC announced that it was restructuring and would henceforth only include trade associations in its membership. While the companies leaving the GCC are still represented by their trade associations, their loss of confidence in the GCC's ability to represent their corporate interests is all too evident.[15]

Thoughtful corporate leaders now know that our energy future is going to be strikingly different from our energy past. There is a growing acceptance among the key energy players that the world is in the early stages of the transition from a carbon-based to a hydrogen-based energy economy. In February 1999, ARCO Chief Executive Officer Michael Bowlin said in a talk at an energy conference in Houston, Texas, "We've embarked on the beginning of the Last Days of the Age of Oil." He went on to discuss the need to convert our carbon-based energy economy into a hydrogen-based energy economy.[16]

Whether the GCC will survive as a collection of trade associations or whether it will join the Tobacco Institute, which closed its doors in January 1999, is uncertain. What is clear is that the organization that so effectively undermined U.S. leadership in Kyoto is no longer a dominant player in the global climate debate. The stage is set for the United States to resume leadership of the global climate stabilization effort.

For additional information, see <www.earth-policy.org/Alerts/Alert6.htm>.

August 2000

Climate Change Has World Skating on Thin Ice

Lester R. Brown

If any explorers had been hiking to the North Pole this summer, they would have had to swim the last few miles. The discovery of open water at the Pole by an ice-breaker cruise ship in mid-August surprised many in the scientific community.[1]

This finding, combined with two recent studies, provides not only more evidence that the Earth's ice cover is melting, but that it is melting at an accelerating rate. A study by two Norwegian scientists projects that within 50 years, the Arctic Ocean could be ice-free during the summer. The other, a study by a team of four U.S. scientists, reports that the vast Greenland ice sheet is melting.[2]

The projection that the Arctic Ocean will lose all its summer ice is not surprising, since an earlier study reported that the thickness of the ice sheet has been reduced by 42 percent over the last four decades. The area of the ice sheet has also shrunk by 6 percent. Together this thinning and shrinkage have reduced the Arctic Ocean ice mass by nearly half.[3]

Meanwhile, Greenland is gaining some ice in the higher altitudes, but it is losing much more at lower elevations, particularly along its southern and eastern coasts. The huge island of 2.2 million square kilometers (three times the size of Texas) is experiencing a net loss of some 51 billion cubic meters of water each year, an amount equal to the annual flow of the Nile River.[4]

The Antarctic is also losing ice. In contrast to the North Pole, which is covered by the Arctic Sea, the South Pole is covered by the Antarctic continent, a land mass roughly the size of the United States. Its continent-sized ice sheet, which is on average 2.3 kilometers (1.5 miles) thick, is relatively stable. But the ice shelves, the portions of the ice sheet that extend into the surrounding seas, are fast disappearing.[5]

A team of U.S. and British scientists reported in 1999 that the ice shelves on either side of the Antarctic Peninsula are in full retreat. From roughly mid-century through 1997, these areas lost 7,000 square kilometers as the ice sheet disintegrated. But then within scarcely a year they lost another 3,000 square kilometers. Delaware-sized icebergs that have broken off are threatening ships in the area. The scientists attribute the accelerated ice melting to a regional temperature rise of some 2.5 degrees Celsius (4.5 degrees Fahrenheit) since 1940.[6]

These are not the only examples of melting. Lisa Mastny, who has reviewed some 30 studies on this topic, reports that ice is melting almost everywhere—and at an accelerating rate. The snow/ice mass is shrinking in the world's major mountain ranges: the Rocky Mountains, the Andes, the Alps, and the Himalayas. In Glacier National Park in Montana, the number of glaciers has dwindled from 150 in 1850 to fewer than 50 today. The U.S. Geological Survey projects that the remaining glaciers will disappear within 30 years.[7]

Scientists studying the Quelccaya glacier in the Peruvian Andes report that its retreat has accelerated from 3 meters a year between roughly 1970 and 1990 to 30 meters a year since 1990. In Europe's Alps, the shrinkage of the glacial area by 35–40 percent since 1850 is expected to continue. These ancient glaciers could largely disappear over the next half-century.[8]

Shrinkage of ice masses in the Himalayas has accelerated alarmingly. In eastern India, the Dokriani Bamak glacier, which retreated by 16 meters between 1992 and 1997, drew back by a further 20 meters in 1998 alone.[9]

This melting and shrinkage of snow/ice masses should not come as a total surprise. Swedish scientist Svante Arrhenius warned at the beginning of the last century that burning fossil fuels could raise atmospheric levels of carbon dioxide (CO_2), creating a greenhouse effect. Atmospheric CO_2 levels, estimated at 280 parts per million (ppm) before the Industrial Revolution, have climbed from 317 ppm in 1960 to 368 ppm in 1999—a gain of 16 percent in only four decades.[10]

As CO_2 concentrations have risen, so too has the earth's temperature. Between 1975 and 1999, the average temperature increased from 13.94 degrees Celsius to 14.35 degrees, a gain of 0.41 degrees or 0.74 degrees Fahrenheit in 24 years. The warmest 23 years since recordkeeping began in 1866 have all occurred since 1975.[11]

Researchers are discovering that a modest rise in temperature of only 1 or 2 degrees Celsius in mountainous regions can dramatically increase the share of precipitation falling as rain while decreasing the share coming down as snow. The result is more flooding during the rainy season, a shrinking snow/ice mass, and less snowmelt to feed rivers during the dry season.[12]

These "reservoirs in the sky," where nature stores fresh water for use in the summer as the snow melts, are shrinking and some could disappear entirely. This will affect the water supply for cities and for irrigation in areas dependent on snowmelt to feed rivers.

If the massive snow/ice mass in the Himalayas—which is the third largest in the world, after the Greenlandic and Antarctic ice sheets—continues to melt, it will affect the water supply of much of Asia. All of the region's major rivers—the Indus, Ganges, Mekong, Yangtze, and Yellow—originate in the Himalayas. The melting in the Himalayas could alter the hydrology of several Asian countries, including Pakistan, India, Bangladesh, Thailand, Viet Nam, and China. Less snowmelt in the summer dry season to feed rivers could exacerbate the hydrological poverty already affecting so many in the region.

As the ice on land melts and flows to the sea, sea level rises. Over the last century, sea level rose by 20–30 centimeters (8–12 inches). During this century, the existing climate models indicate it could rise by as much as 1 meter. If the Greenland ice sheet, which is up to 3.2 kilometers thick in places, were to melt entirely, sea level would rise by 7 meters (23 feet).[13]

Even a much more modest rise would affect the low-lying river floodplains of Asia, where much of the region's rice is produced. According to a World Bank analysis, a 1-meter rise in sea level would cost low-lying Bangladesh half its riceland. Numerous low-lying island countries would have to be evacuated. The residents of densely populated river valleys of Asia would be forced inland into already crowded interiors. Rising sea level could create climate refugees by the million in countries such as China, India, Bangladesh, Indonesia, Viet Nam, and the Philippines.[14]

Even more disturbing, ice melting itself can accelerate temperature rise. As snow/ice masses shrink, less sunlight is reflected back into space. With more sunlight absorbed by less reflective surfaces, temperature rises even faster and melting accelerates.

We don't have to sit idly by as this scenario unfolds. There may still be time to stabilize atmospheric CO_2 levels before continuing carbon emissions cause climate change to spiral out of control. We have more than enough wind, solar, and geothermal energy that can be economically harnessed to power the world economy. If we were to incorporate the cost of climate disruption in the price of fossil fuels in the form of a carbon tax, investment would quickly shift from fossil fuels to these climate-benign energy sources.

The leading automobile companies are all working on fuel-cell engines. DaimlerChrysler plans to start marketing such an automobile in 2003. The fuel of choice for these engines is hydrogen. Even leaders within the oil industry recognize that we will eventually shift from a carbon-based energy economy to a hydrogen-based one. The question is whether we can make that shift before the earth's climate system is irrevocably altered.[15]

For additional information, see <www.earth-policy.org/Alerts/Alert7.htm>.

September 2000

OPEC Has World Over a Barrel Again

Lester R. Brown

On Thursday, September 7, 2000, oil prices on the spot market climbed to $35.39 per barrel, their highest since November 1990, just before the Gulf War. This latest oil price escalation not only threatens a worldwide recession, it also marks another adverse shift in the international terms of trade for the United States, one that will widen further the already huge trade deficit.[1]

On Sunday, OPEC (Organization of Petroleum Exporting Countries) ministers will meet at OPEC headquarters in Vienna to consider a request from oil-importing countries to boost daily oil output by at least 500,000 barrels. But it may be too little too late. With the East Asian economies, including that of China, booming again, and with U.S. oil production falling for eight years in a row, even a production increase of 500,000 barrels may not restore lower oil prices.[2]

For the United States, which pays for its oil imports in part with grain exports, this is not good news. Exports of grain and oil are each concentrated in a handful of countries, with grain coming largely from North America and oil mostly from the Middle East. The United States, which dominates grain exports even more than Saudi Arabia does oil, is both the world's leading grain exporter and its biggest oil importer. Ironically, all 11 members of OPEC are grain importers.[3]

Using the price of wheat as a surrogate for grain prices,

shifts in the grain/oil exchange rate can be easily monitored. (See Table 3–1.) From 1950 through 1972, both wheat and oil prices were remarkably stable. In 1950, when wheat was priced at $1.89 a bushel and oil at $1.71 a barrel, a bushel of wheat could be exchanged for 1.1 barrels of oil. At any time during this 22-year span, a bushel of wheat could be traded for a barrel of oil on the world market.[4]

With the 1973 oil price hike, this began to change. By 1979, the year of the second oil price increase, OPEC's strength had pushed the exchange rate to roughly 4 to 1. By 1982, when the price of oil had climbed past $33 a barrel, the wheat/oil ratio had climbed to 8 to 1. This steep rise in the purchasing power of oil led to one of the greatest international transfers of wealth ever recorded.[5]

Today, 27 years after the first oil price hike, the terms of trade are again shifting in favor of OPEC. With grain prices at their lowest level in two decades and oil prices at the highest level in a decade, the wheat/oil ratio has shifted to an estimated 10 to 1 this year. OPEC has the United States over a barrel once again. With its fast-growing fleet of gas-guzzling SUVs (sport utility vehicles) and falling oil production, the United States is now dependent on imports for a record 57 percent of its oil, making it even more vulnerable to oil price hikes and supply disruptions than it was in 1973.[6]

But this is not the only threat to international security. Climate change from burning oil and other fossil fuels may be an even greater threat to long-term world economic and political stability. Last month's discovery of open water at the North Pole by an ice breaker cruise ship is only one of many recent indications that human activities are altering the earth's climate. The Arctic Ocean ice has thinned by 40 percent in some 35 years. Scientists now believe that summer ice in the Arctic Ocean could disappear entirely within the next 50 years.[7]

Table 3–1. *The Wheat-Oil Exchange Rate, 1950–2000*

Year	Bushel of Wheat	Barrel of Oil	Bushels Per Barrel
	(U.S. dollars[1])		(Ratio)
1950	1.89	1.71	1
1955	1.81	1.93	1
1960	1.58	1.50	1
1965	1.62	1.33	1
1970	1.49	1.30	1
1971	1.68	1.65	1
1972	1.90	1.90	1
1973	3.81	2.70	1
1974	4.89	9.76	2
1975	4.06	10.72	3
1976	3.62	11.51	3
1977	2.81	12.40	4
1978	3.48	12.70	4
1979	4.36	17.26	4
1980	4.70	28.67	6
1981	4.76	32.50	7
1982	4.36	33.47	8
1983	4.28	29.31	7
1984	4.15	28.25	7
1985	3.70	26.98	7
1986	3.13	13.82	4
1987	3.07	17.79	6
1988	3.95	14.15	4
1989	4.61	17.19	4
1990	3.69	22.05	6
1991	3.50	18.30	5

Table 3–1. *continued*

1992	4.11	18.22	4
1993	3.82	16.13	4
1994	4.08	15.47	4
1995	4.82	17.20	4
1996	5.64	20.37	4
1997	4.35	19.27	4
1998	3.43	13.07	4
1999	3.05	17.98	6
2000 (est.)	2.94	29.34	10

[1]Prices in current year dollars.
Source: International Monetary Fund, *International Financial Statistics* (Washington, DC: various years).

Greenland's ice sheet is also starting to melt. If all the ice on this huge island, which is three times the size of Texas and measures over 3,000 meters thick (10,000 feet) in some places, were eventually to melt, sea level would rise by a staggering 7 meters (23 feet). In addition to ice melting and rising sea level, global climate change can bring more extreme weather events—more intense heat waves, more destructive storms, and more severe flooding.[8]

The world is beginning to move beyond oil and coal toward energy sources that do not disrupt climate. Widely varying growth rates of various sources of energy from 1990–99 give a sense of the energy transition under way. Worldwide, wind power generation grew by 24 percent per year, solar cell production by 17 percent, and geothermal power by 4 percent. By contrast, world oil use expanded at 1 percent a year and coal use actually declined by nearly 1 percent.[9]

Even oil company CEOs are talking about shifting from a carbon-based to a solar/hydrogen-based energy economy. British Petroleum is now the world's leading manufacturer of solar cells. Shell is pioneering the new hydrogen economy. All the major automobile companies are working on fuel-cell engines for which the fuel of choice is hydrogen. The Japanese have developed a photovoltaic roofing material that allows the rooftop to become the power plant for the building.[10]

Denmark now gets 10 percent of its electricity from wind. For Schleswig-Holstein, the northernmost state in Germany, it is 14 percent. For the industrial province of Navarra in Spain, it is 22 percent. We are now getting glimpses of the new energy economy in the solar rooftops in Japan and in the wind turbines scattered across the European countryside.[11]

A nationwide wind resources survey by the U.S. Department of Energy indicates that three states—Kansas, North Dakota, and Texas—have enough harnessable wind energy to satisfy national electricity needs. With new wind farms coming online over the last year or two in Iowa, Minnesota, Texas, and Wyoming, U.S. wind-generation jumped by 29 percent in 1999.[12]

The generation of electricity from wind is exciting because money spent for this electricity typically stays in the community, whereas money spent for electricity generated by oil may end up in the Middle East. Moreover, with cheap wind-generated electricity, hydrogen, the preferred fuel for fuel-cell engines, can be produced during the night when electricity demand is low.

As these examples indicate, the transition to a new energy economy has begun, but it is not moving fast enough. The time has come to restructure the tax system both to reduce the threat of soaring oil prices and to stabilize climate. We can restructure our tax system by lowering the personal and corporate income tax and offsetting it with an increase in a tax on gasoline. OPEC members know that the cost of producing oil in Saudi Arabia, which has the lion's share of world oil reserves, is roughly $2 a barrel. They also know that if they push the price of oil too high, they will trigger a global recession. This is not in their interest.[13]

If there is a world price for petroleum products beyond which a further rise would be disruptive, then the issue is who gets the difference between the low production cost of oil and this much higher market price. If importing countries push prices of gasoline, fuel oil, jet fuel, and other oil products close to that limit by imposing stiff taxes, then the potential for raising prices by OPEC is lessened. This is why, in a meeting with President Clinton in New York earlier this week, Saudi Crown Prince Abdullah urged importing countries to lower their taxes on gasoline and other oil products.[14]

If we take the initiative and raise gasoline taxes while lowering income taxes, the increase in the gasoline tax will end up in our treasury and individuals will benefit from lower income taxes. But if we don't restructure and let OPEC countries keep increasing the price of oil, and hence of gasoline, the equivalent of the gasoline tax increase will end up in OPEC treasuries. We will eventually pay the same higher price for gasoline, but not get the income tax reduction.

For additional information, see <www.earth-policy.org/Alerts/Alert8.htm>.

May 2001

Wind Power: The Missing Link in the Bush Energy Plan

Lester R. Brown

The eagerly awaited Bush energy plan released on May 17, 2001, disappointed many people because it largely overlooked the potential contribution of raising energy efficiency. It also overlooked the enormous potential of wind power, which is likely to add more to U.S. generating capacity over the next 20 years than coal.[1]

In short, the authors of the plan appear to be out of touch with what is happening in the world energy economy, fashioning an energy plan more appropriate for the early twentieth century rather than the early twenty-first century. They emphasized the role of coal, but world coal use peaked in 1996 and has declined some 11 percent since then as countries have turned away from this climate-disrupting fuel. Even China, which rivals the United States as a coal burning country, has reduced its coal use by 24 percent since 1996.[2]

Meanwhile, world wind power use has multiplied nearly fourfold over the last five years, a growth rate matched only by the computer industry. In the United States, the American Wind Energy Association projects a staggering 60-percent growth in wind-generating capacity this year.[3]

Wind power was once confined to California, but during the last three years, wind farms coming online in Minnesota, Iowa, Texas, Colorado, Wyoming, Oregon, and Pennsylvania have boosted U.S. capacity by half from 1,680 megawatts to 2,550 megawatts. The 1,500 or more megawatts to be added this year will be located in a dozen states. A 300-megawatt wind farm under construction on the Oregon/Washington border is currently the world's largest.[4]

But this is only the beginning. The Bonneville Power Administration (BPA) indicated in February that it wanted to buy 1,000 megawatts of wind-generating capacity and requested proposals. Much to its surprise, it received enough to build 2,600 megawatts of capacity in five states, with the potential of expanding these sites to over 4,000 megawatts. BPA, which may accept most of these proposals, expects to have at least one site online by the end of this year.[5]

A 3,000-megawatt wind farm in the early planning stages in South Dakota, near the Iowa border, is 10 times the size of the Oregon/Washington wind farm. Named Rolling Thunder, this project, initiated by Dehlsen Associates and drawing on the leadership of Jim Dehlsen, a wind energy pioneer in California, is designed to feed power to the midwestern region around Chicago. This proposed project is not only large by wind power standards, it is one of the largest energy projects of any kind in the world today.[6]

Advances in wind turbine technology, drawing heavily from the aerospace industry, have lowered the cost of wind power from 38¢ per kilowatt-hour in the early 1980s to 3–6¢ today, depending on the wind site. Wind, now competitive with fossil fuels, is already cheaper in some locations than oil or gas-fired power. With major corporations such as ABB, Shell International, and Enron plowing resources into this field, further cost cuts are in prospect.[7]

Wind is a vast, worldwide source of energy. The U.S.

Great Plains are the Saudi Arabia of wind power. Three wind-rich U.S. states—Kansas, North Dakota, and Texas—have enough harnessable wind to meet national electricity needs. China can double its existing generating capacity from wind alone. Densely populated Western Europe can meet all of its electricity needs from offshore wind power.[8]

Today Denmark, the world leader in wind turbine technology and manufacture, is getting 15 percent of its electricity from wind power. For Schleswig-Holstein, the northernmost state of Germany, the figure is 19 percent, and for some parts of the state, 75 percent. Spain's industrial province of Navarra, starting from scratch six years ago, now gets 22 percent of its electricity from wind.[9]

As wind-generating costs fall and as concern about climate change escalates, more and more countries are climbing onto the wind energy bandwagon. In December, France announced it will develop 5,000 megawatts of wind power by 2010. Also in December, Argentina announced a plan to develop 3,000 megawatts of wind power in Patagonia by 2010. In April, the United Kingdom accepted offshore bids for 1,500 megawatts of wind power.[10]

The growth in wind power is consistently outrunning earlier estimates. The European Wind Energy Association, which in 1996 had set a target of 40,000 megawatts for Europe in 2010, recently upped it to 60,000 megawatts.[11]

The Bush plan to add 393,000 megawatts of electricity nationwide by 2020 could be satisfied from wind alone. Money spent on wind-generated electricity tends to remain in the community, providing income, jobs, and tax revenue, bolstering local economies. One large advanced-design wind turbine, occupying a quarter-acre of land, can easily yield a farmer or rancher $2,000 in royalties per year while providing the community with $100,000 of electricity. U.S. farmers and ranchers, who own most of the wind rights in the country, are now joining environmentalists to lobby for development of this abundant alternative to fossil fuel.[12]

Once we get cheap electricity from wind, we can use it to electrolyze water, producing hydrogen. Hydrogen is the fuel of choice for the new, highly efficient, fuel-cell engine that every major automobile manufacturer is now working on. DaimlerChrysler plans to be on the market with fuel cell-powered cars in 2003. Ford, Toyota, and Honda will probably not be far behind. William Ford, Chairman of Ford Motor Company, says he expects to preside over the demise of the internal combustion engine.[13]

Surplus wind power can be stored as hydrogen and used in fuel cells or gas turbines to generate electricity, leveling supply when winds are variable. Wind, once seen as a cornerstone of the new energy economy, may turn out to be its foundation. The wind meteorologist who analyzes wind regimes and identifies the best sites for wind farms will play a role in the new energy economy comparable to that of the petroleum geologist in the old energy economy.

With the advancing technologies for harnessing wind and powering motor vehicles with hydrogen, we can now see a future where farmers and ranchers can supply not only much of the country's electricity, but much of the hydrogen to fuel its fleet of automobiles as well. For the first time, the United States has the technology and resources to divorce itself from Middle Eastern oil.

In addition to neglecting the potential of wind, the Bush energy strategy pays only lip service to climate stabilization. This is a high-risk strategy. With business as usual, the International Panel on Climate Change recently projected a global temperature rise during this century of up to 6 degrees Celsius (10 degrees Fahrenheit). If this rise occurs, the rest of the world may hold the United States, the leading carbon emitter, responsible.[14]

What the United States needs now is an energy plan for this century, one that takes into account not only recent technological advances in wind power, fuel cells, and hydrogen generators, but also the need to stabilize climate. Perhaps Congress will bring the energy plan into the twenty-first century and restore U.S. leadership in the fast-changing world energy economy.

For additional information, see <www.earth-policy.org/Alerts/Alert14.htm>.

Population and Health

June 2000

Population Growth Sentencing Millions to Hydrological Poverty

Lester R. Brown

At a time when drought in the United States, Ethiopia, and Afghanistan is in the news, it is easy to forget that far more serious water shortages are emerging as the demand for water in many countries simply outruns the supply. Water tables are now falling on every continent. Literally scores of countries are facing water shortages as water tables fall and wells go dry.

We live in a water-challenged world, one that is becoming more so each year as 80 million additional people stake their claims to the earth's water resources. Unfortunately, nearly all the projected 3 billion people to be added over the next half-century will be born in countries that are already experiencing water shortages. Even now many in these countries lack enough water to drink, to satisfy hygienic needs, and to produce food.[1]

By 2050, India is projected to add 519 million people and China 211 million. Pakistan is projected to add nearly 200 million, going from 151 million at present to 348 million. Egypt, Iran, and Mexico are slated to increase their populations by more than half by 2050. In these and other water-short countries, population growth is sentencing millions of people to hydrological poverty, a local form of poverty that is difficult to escape.[2]

Even with today's 6 billion people, the world has a huge water deficit. Using data on overpumping for China, India, Saudi Arabia, North Africa, and the United States, Sandra Postel, author of *Pillar of Sand: Can the Irrigation Miracle Last?*, calculates the annual depletion of aquifers at 160 billion cubic meters or 160 billion tons. Using the rule of thumb that it takes 1,000 tons of water to produce 1 ton of grain, this 160-billion-ton water deficit is equal to 160 million tons of grain or one half the U.S. grain harvest.[3]

At average world grain consumption of just over 300 kilograms or one third of a ton per person per year, this would feed 480 million people. Stated otherwise, 480 million of the world's 6 billion people are being fed with grain produced with the unsustainable use of water.[4]

Overpumping is a new phenomenon, one largely confined to the last half-century. Only since the development of powerful diesel and electrically driven pumps have we had the capacity to pull water out of aquifers faster than it is replaced by precipitation.

Some 70 percent of the water consumed worldwide, including both that diverted from rivers and that pumped from underground, is used for irrigation, while some 20 percent is used by industry, and 10 percent for residential purposes. In the increasingly intense competition for water among sectors, agriculture almost always loses. The 1,000 tons of water used in India to produce 1 ton of wheat worth perhaps $200 can also be used to expand industrial output by easily $10,000, or 50 times as much. This ratio helps explain why, in the American West, the sale of irrigation water rights by farmers to cities is an almost daily occurrence.[5]

In addition to population growth, urbanization and industrialization also expand the demand for water. As developing-country villagers, traditionally reliant on the village well, move to urban high-rise apartment buildings with indoor plumbing, their residential water use can easily triple. Industrialization takes even more water than urbanization.

Rising affluence in itself generates additional demand for water. As people move up the food chain, consuming more beef, pork, poultry, eggs, and dairy products, they use more grain. A U.S. diet rich in livestock products requires 800 kilograms of grain per person a year, whereas diets in India, dominated by a starchy food staple such as rice, typically need only 200 kilograms. Using four times as much grain per person means using four times as much water.[6]

Once a localized phenomenon, water scarcity is now crossing national borders via the international grain trade. The world's fastest-growing grain import market is North Africa and the Middle East, an area that includes Morocco, Algeria, Tunisia, Libya, Egypt, and the Middle East through Iran. Virtually every country in this region is simultaneously experiencing water shortages and rapid population growth.[7]

As the demand for water in the region's cities and industries increases, it is typically satisfied by diverting water from irrigation. The loss in food production capacity is then offset by importing grain from abroad. Since 1 ton of grain represents 1,000 tons of water, this becomes the most efficient way to import water.[8]

Last year, Iran imported 7 million tons of wheat, eclipsing Japan to become the world's leading wheat importer. This year, Egypt is also projected to move ahead of Japan. Iran and Egypt have nearly 70 million people each. Both populations are increasing by more than a million a year and both are pressing against the limits of their water supplies.[9]

The water required to produce the grain and other foodstuffs imported into North Africa and the Middle East last year was roughly equal to the annual flow of the Nile River. Stated otherwise, the fast-growing water deficit of this region is equal to another Nile flowing into the region in the form of imported grain.[10]

It is now often said that future wars in the region will more likely be fought over water than oil. Perhaps, but given the difficulty in winning a water war, the competition for water seems more likely to take place in world grain markets. The countries that will "win" in this competition will be those that are financially strongest, not those that are militarily strongest.

The world water deficit grows larger with each year, making it potentially more difficult to manage. If we decided abruptly to stabilize water tables everywhere by simply pumping less water, the world grain harvest would fall by some 160 million tons, or 8 percent, and grain prices would go off the top of the chart. If the deficit continues to widen, the eventual adjustment will be even greater.[11]

Unless governments in water-short countries act quickly to stabilize population and to raise water productivity, their water shortages may soon become food shortages. The risk is that the growing number of water-short countries, including population giants China and India, with rising grain import needs will overwhelm the exportable supply in grain-surplus countries, such as the United States, Canada, and Australia. This in turn could destabilize world grain markets.

Another risk of delay in dealing with the deficit is that

some low-income, water-short countries will not be able to afford to import needed grain, trapping millions of their people in hydrological poverty, thirsty and hungry, unable to escape.

Although there are still some opportunities for developing new water resources, restoring the balance between water use and the sustainable supply will depend primarily on demand-side initiatives, such as stabilizing population and raising water productivity.

Governments can no longer separate population policy from the supply of water. And just as the world turned to raising land productivity a half-century ago when the frontiers of agricultural settlement disappeared, so it must now turn to raising water productivity. The first step toward this goal is to eliminate the water subsidies that foster inefficiency. The second step is to raise the price of water to reflect its cost. Shifting to more water-efficient technologies, more water-efficient crops, and more water-efficient forms of animal protein offers a huge potential for raising water productivity. The shifts will move faster if the price of water more closely reflects its value.

For additional information, see <www.earth-policy.org/Alerts/Alert4.htm>.

July 2000

Africa Is Dying—It Needs Help

Lester R. Brown

The recent International AIDS conference in Durban, South Africa, reminds us that Africa is dying. The HIV epidemic that is raging across Africa is now taking some 6,030 lives each day, the equivalent of 15 fully loaded jumbo jets crashing—with no survivors. This number, climbing higher each year, is expected to double during this decade.[1]

Public attention has initially focused on the dramatic rise in adult mortality and the precipitous drop in life expectancy. But we need now to look at the longer term economic consequences—falling food production, deteriorating health care, and disintegrating educational systems. Effectively dealing with this epidemic and the heavy loss of adults will make the rebuilding of Europe after World War II seem like child's play by comparison.

While industrial countries have held the HIV infection rate among the adult population to less than 1 percent, in some 16 African countries it is over 10 percent. (See Table 3–2.) In South Africa, it is 20 percent. In Zimbabwe and Swaziland, it is 25 percent. And in Botswana, which has the highest infection rate, 36 percent of adults are HIV-positive. Barring a medical miracle, these latter countries will lose one fifth to one third of their adults by the end of this decade.[2]

Attention in Durban focused on the high cost of treating those already ill, but the virus is continuing to spread.

Unless its spread is curbed soon, it will take more lives in Africa than World War II claimed worldwide.

As deaths multiply, life expectancy falls. Without AIDS, countries with high infection rates, like Botswana, Zimbabwe, and South Africa would have a life expectancy of some 70 years or more. With the virus continuing to spread, life expectancy could drop to 30—more like a medieval life span than a modern one.[3]

Whereas infectious diseases typically take their heaviest toll among the eldest and the very young who have weaker immune systems, HIV claims mostly adults, depriving countries of their most productive workers. In the epidemic's early stages, the virus typically spreads most rapidly among the better educated, more socially mobile segment of society. It takes the agronomists, engineers, and teachers on whom economic development depends.

The HIV epidemic is affecting every segment of society, every sector of the economy, and every facet of life. For example, close to half of Zimbabwe's health care budget is used to treat AIDS patients. In some hospitals in Burundi and South Africa, AIDS patients occupy 60 percent of the beds. Health care workers are worked to exhaustion.[4]

This epidemic, now producing thousands of orphans each day, could easily produce 40 million orphans by 2010, a number that could overwhelm the resources of extended families.[5]

Education is also suffering. In Zambia, the number of teachers dying with AIDS each year approaches the number of new teachers being trained. In the Central African Republic, a shortage of teachers closed 107 primary schools, leaving only 66 open. At the college level, the damage is equally devastating. At the University of Durban-Westville in South Africa, 25 percent of the student body is HIV-positive.[6]

Table 3–2. *Countries Where HIV Infection Rate Among Adults Is Greater Than 10 Percent*

Country	Population	Share of Adult Population Infected
	(million)	(percent)
Botswana	2	36
Swaziland	1	25
Zimbabwe	12	25
Lesotho	2	24
South Africa	40	20
Zambia	9	20
Namibia	2	20
Malawi	11	16
Kenya	30	14
Central African Republic	4	14
Mozambique	19	13
Côte d'Ivoire	14	12
Djibouti	1	12
Burundi	7	11
Rwanda	7	11
Ethiopia	61	11

Source: UNAIDS, *Report on the Global HIV/AIDS Epidemic* (Geneva: June 2000).

In addition to the continuing handicaps of a lack of infrastructure and trained personnel, Africa must now contend with the adverse economic effects of the epidemic. AIDS dramatically increases the dependency ratio, the number of young and elderly who depend on productive adults. This in turn makes it much more difficult for a society to save. Reduced savings means reduced investment and slower economic growth or even decline.

At the corporate level, firms in countries with high infection rates are seeing their employee health care insurance costs double, triple, or quadruple. Companies that were until recently comfortably in the black now find themselves in the red. Under these circumstances, investment inflows from abroad are declining and could dry up entirely.

In a largely rural society, food security declines as the epidemic progresses. At the family level, food supplies drop precipitously when the first adult develops full-blown AIDS. This deprives the family not only of this worker in the fields, but also of the work time of the adult caring for the AIDS victim. A survey in Tanzania found that a woman whose husband was sick with AIDS spent 60 percent less time tending the crops.[7]

Food production declines from the epidemic have been reported in Burkina Faso, Côte d'Ivoire, and Zimbabwe. In pastoral economies, such as Namibia, the loss of the male head of household is often followed by the loss of cattle, the family's livelihood.[8]

Sub-Saharan Africa, a region of 600 million people, is moving into uncharted territory. There are historical precedents for epidemics on this scale, such as the smallpox epidemic that decimated New World Indian populations in the sixteenth century or the bubonic plague in Europe in the fourteenth century, but there is no precedent for such a concentrated loss of adults.[9]

The good news is that some countries are halting the spread of the virus. The key is strong leadership from the top. In Uganda, where the epidemic first took root, the active personal leadership of President Yoweri Museveni over the last dozen years has succeeded in reducing the share of adults infected with the virus from a peak of 14 percent to 8 percent. In effect, the number of new infections has dropped well below the number of deaths from AIDS.[10]

Senegal, alone in Africa, responded early to the threat from the virus. As a result, it prevented the epidemic from gaining momentum and held the infection rate to 2 percent of its adults, a number only slightly higher than that of the industrial countries.[11]

Saving Africa depends on a Marshall Plan–scale effort on two fronts: one to curb the spread of the virus and the other to restore economic progress. Winning the former depends directly on Africa's national political leaders. Unless they personally lead, the effort will fail.

Once the leader outlines the behavioral changes needed to contain the virus—such as young people delaying first intercourse, reducing the number of sexual partners, and using condoms—then others can contribute. This includes the medical establishment within the country, nongovernmental groups working in this area, and international health and family planning agencies.

To compensate for the "missing generation," countries will need assistance across the board in education. This is an area where the U.S. Peace Corps and its equivalents in Europe can play a central role, particularly in supplying the teachers needed to keep schools open. Social workers are needed to work with orphans. A program of financial assistance is necessary for the extended families trying to absorb the millions of orphans projected by 2010.

Given the high cost of doing business in an AIDS-ridden society, special incentives in the form of tax relief are needed to attract corporate investors, incentives that could be underwritten by international development agencies. And it goes without saying, debt relief is essential to the rebuilding of Africa.

It is not possible to outline a detailed rescue effort here. The bottom line is that there is no precedent in international development for the challenge the world now faces in Africa. The question is not whether we can respond to this challenge. We can. We have the resources to do so. If we fail to respond to Africa's pain, we will forfeit the right to call ourselves a civilized society.

For additional information, see <www.earth-policy.org/Alerts/Alert5.htm>.

October 2000

HIV Epidemic Restructuring Africa's Population

Lester R. Brown

The HIV epidemic raging across Africa is a tragedy of epic proportions, one that is altering the region's demographic future. It is reducing life expectancy, raising mortality, lowering fertility, creating an excess of men over women, and leaving millions of orphans in its wake.

This year began with 24 million Africans infected with the virus. In the absence of a medical miracle, nearly all will die before 2010. Each day, 6,000 Africans die from AIDS. Each day, an additional 11,000 are infected.[1]

The epidemic has proceeded much faster in some countries than in others. In Botswana, 36 percent of the adult population is HIV-positive. In Zimbabwe and Swaziland, the infection rate is 25 percent. Lesotho is at 24 percent. In Namibia, South Africa, and Zambia, the figure is 20 percent. In none of these countries has the spread of the virus been checked.[2]

Life expectancy, a sentinel indicator of economic progress, is falling precipitously. In Zimbabwe, without AIDS, life expectancy in 2010 would be 70 years, but with AIDS, it is expected to fall below 35 years. Botswana's life expectancy is projected to fall from 66 years to 33 years by 2010. For South Africa, it will fall from 68 years to 48 years. And for Zambia, from 60 to 30 years. These life expectancies are more akin to those of the Middle Ages than of the modern age.[3]

The demography of this epidemic is not well understood simply because, in contrast to most infectious diseases, which take their heaviest toll among the elderly and the very young, this virus takes its greatest toll among young adults. The effect on mortality is most easily understood. In the absence of a low-cost cure, infection leads to death. The time from infection until death for adults in Africa is estimated at 7 to 10 years.[4]

This means that Botswana can expect to lose the 36 percent of its adult population that is HIV-positive within this decade, plus the additional numbers who will be infected within the next year or two. The HIV toll, plus normal deaths among adults, means that close to half of the adults in Botswana today will be dead by 2010. Other countries with high infection rates, such as South Africa, Swaziland, and Zimbabwe, will likely lose nearly a third of their adults by 2010.[5]

Adults are not the only ones dying from AIDS. In Africa, infants of mothers who are HIV-positive have a 30- to 60-percent chance of being born with the virus. Their life expectancy is typically less than two years. Many more infants acquire the virus through breastfeeding. Few of them will reach school age.[6]

Thus far, attention has focused on the effect of rising mortality on future population trends, but the virus also reduces fertility. Research is limited, but early evidence indicates that from the time of infection onward, fertility among infected women slowly declines. By the time symptoms of AIDS appear, women are 70 percent less likely to be pregnant than those who are not infected.[7]

Females are infected at an earlier age than males because they have sexual relations with older men who are more likely to be HIV-positive. Female infection rates are also higher than those of males. Among 15- to 19-year-olds, five times as many females are infected as males. Because they are infected so early in life, many women will die before completing their reproductive years, further reducing births.[8]

A demographically detailed study in Kisumu, Kenya, found that 8 percent of 15-year-old girls are HIV-positive. For 16-year-olds, the figure is 18 percent; and by age 19, it is 33 percent. Among the 19-year-olds, the average age of infection was roughly 17 years. With a life expectancy of perhaps nine years after infection, the average woman in this group will die at age 26, long before her child-bearing years are over.[9]

Much work remains to be done in analyzing the effects of the HIV epidemic on fertility, but we do know that with other social traumas, such as famine, the effect of fertility decline on population size can equal the effect of rising mortality. For example, in the 1959–61 famine in China, some 30 million Chinese starved to death, but the actual reduction in China's population as a result of the famine was closer to 60 million.[10]

The reasons are well understood. In a famished population, the level of sexual activity declines, many women stop ovulating, and even the women who do conceive often abort spontaneously. In a prolonged famine, the fall in births can contribute as much to the population decline as the rise in mortality. How much the HIV epidemic will eventually reduce fertility no one knows.

One thing is known: The wholesale death of young adults in Africa is creating millions of orphans. By 2010, Africa is expected to have 40 million orphans. Although Africa's extended family system is highly resilient and capable of caring for children left alone when parents die, it will be staggered by this challenge. There is a real possibility that millions of orphans will become street children, trying to survive by whatever means they can.[11]

Africa is also facing a gender imbalance, a unique shortage of women. After wars, countries often face a severe shortage of males, as Russia did after World War II. This epidemic, however, is claiming more females than males in Africa, promising a future where men will outnumber women 11 to 9. This will leave many males either destined to bachelorhood or forced to migrate to countries outside the region in search of a wife.[12]

The demographic effects of the HIV epidemic on Africa will be visible for generations to come. Until recently, the official projections at the United Nations indicated continuing population growth in all countries in Africa. Now this may be changing as the United Nations acknowledges that populations could decline in some countries. If the new U.N. biennial update of world population numbers and projections, due out before the end of this year, includes the full effect of the epidemic on fertility as well as on mortality, it will likely show future population declines for many African countries, including Botswana, Zimbabwe, South Africa, and Zambia.

There are many unknowns in the effects of the HIV epidemic on the demographic equation. Will health care systems, overwhelmed by AIDS victims, be able to meet the need for basic health care? How will the loss of so many adults in rural communities affect food security? What will be the effect on fertility of women surrounded by death? What will be the social effects of the missing generation of young adults unable to rear their children or to care for their parents?

Even though the HIV epidemic may claim more lives in Africa than World War II claimed worldwide, the epidemic is simply not being given the priority it deserves either within the countries most affected or within the international community. The challenge is to reduce the number of new infections as rapidly as possible. Nothing should deter societies from this goal.

One of the earliest countries hit by the epidemic,

Uganda, has become a model for other countries as the infected share of its adult population has dropped from 14 percent in the early 1990s to 8 percent in 2000, a dramatic achievement. In Zambia, which has mobilized the health, education, agricultural, and industrial sectors, plus church groups, in the effort to curb the spread of the virus, the infected share of young females in some cities has dropped by nearly half since 1993. Zambia may soon turn the HIV tide. If all African countries can do what Uganda has done and what Zambia appears to be doing—namely, reduce the number of new infections below that of AIDS deaths—they may set the stage for ending this history-altering epidemic.[13]

For additional information, see <www.earth-policy.org/Alerts/Alert10.htm>.

December 2000

Obesity Threatens Health in Exercise-Deprived Societies

Lester R. Brown

Obesity is reaching epidemic proportions, afflicting a growing number of people in industrial and developing countries alike. It is damaging human health, raising the incidence of heart disease, stroke, breast cancer, colon cancer, arthritis, and adult onset diabetes. In the United States, the Centers for Disease Control and Prevention estimates that 300,000 Americans now die each year from obesity-related illnesses.[1]

Reducing obesity has traditionally focused on lowering caloric intake by dieting, but there is growing evidence that exercise deprivation is also a major contributor to obesity. With metabolic systems shaped by 4 million years of highly active hunting and gathering, many people may not be able to maintain a healthy body weight without regular exercise.

For the first time in history, a majority of adults in some societies are overweight. In the United States, 61 percent of all adults are overweight. In Russia, the figure is 54 percent; in the United Kingdom, 51 percent; and in Germany, 50 percent. For Europe as a whole, more than half of those between 35 and 65 years of age are overweight.[2]

The number who are overweight is rising in developing countries as well. In Brazil, for example, 36 percent of the adult population is overweight. Fifteen percent of China's adult population is overweight.[3]

Not only are more people overweight than ever before, but their ranks are expanding at a record rate. In the United States, obesity among adults increased by half between 1980 and 1994. Among Americans, 20 percent of men and 25 percent of women are more than 30 pounds (13.6 kilograms) overweight. Surveys in China showed that during the boom years between 1989 and 1992, the share of adults overweight jumped from 9 percent to 15 percent.[4]

Juvenile obesity is rising rapidly. In the United States, where at least 1 out of 10 youngsters 6 to 17 years of age is overweight, the incidence of obesity among children has more than doubled over the last 30 years. Not only does juvenile obesity typically translate into adult obesity, but it also causes metabolic changes that make the disease difficult to treat in adulthood.[5]

Obesity is concentrated in cities. As societies urbanize and people adopt sedentary lifestyles, obesity increases. In both China and Indonesia, the share of people who are obese in cities is double that in the countryside. In the Congo, obesity is six times higher in cities.[6]

In a Worldwatch Paper, *Underfed and Overfed*, Gary Gardner and Brian Halweil report that the number who are overnourished and overweight has climbed to 1.1 billion worldwide, rivaling the number who are undernourished and underweight. Peter Kopelman of the Royal London School of Medicine summarizes the thinking of the medical community: "Obesity should no longer be regarded simply as a cosmetic problem affecting certain individuals, but [as] an epidemic that threatens global well being."[7]

Damage to health from obesity takes many forms. In addition to the illnesses noted earlier, heavier body weight increases resistance to the heart's pumping of blood, elevating blood pressure. It also raises the stress on joints, often causing lower back pain. Those who are obese are four times as likely to have diabetes as those who are not.[8]

As weight goes up, life expectancy goes down. In analyzing this relationship for Americans between the ages of 30 and 42, one broad-based study found that the risk of death within 26 years increased by 1 percent with each additional pound (0.45 kg) of weight.[9]

The estimated 300,000 Americans who die prematurely each year as a result of being overweight is nearing the 400,000 who die prematurely from cigarette smoking. But there is one difference. The number of cigarettes smoked per person in the United States is on the decline, falling some 42 percent between 1980 and 1999, whereas obesity is on the rise. If recent trends continue, it is only a matter of time before deaths from obesity-related illnesses overtake those related to smoking.[10]

Gaining weight is a result of consuming more calories than are burned. With modernization, caloric intake has climbed. Over the last two decades, caloric intake in the United States has risen nearly 10 percent for men and 7 percent for women. Modern diets are rich in fat and sugar. In addition to sugars that occur naturally in food, the average American diet now includes 20 teaspoons of added sugar a day, much of it in soft drinks and prepared foods. Unfortunately, diets in developing countries, especially in urban areas, are moving in this same direction.[11]

While caloric intake has been rising, exercise has been declining. The latest U.S. survey shows that 57 percent of Americans exercise only occasionally or not at all, a number that corresponds closely with the share of the population that is overweight.[12]

Economic modernization has systematically eliminated exercise from our lives. Workers commute by car from home to work in an office or factory, driving quite literally from door to door. Automobiles have eliminated daily walking and cycling. Elevators and escalators have replaced stairs. Leisure time is spent watching television. In the United Kingdom, the two lifestyle variables that correlate most closely with obesity are television viewing and automobile ownership.[13]

Children who watch television five or more hours a day are five times as likely to be overweight as those who watch less than two hours a day. Time spent playing computer games and surfing the Internet in lieu of playing outside is also contributing to the surge in obesity.[14]

A common impulse of those who are overweight is to go on a diet of some sort, attempting to reduce caloric intake to the level of caloric use. Unfortunately, this is physiologically difficult given the abnormally low calorie use associated with our sedentary lifestyles. Ninety-five percent of Americans who attempt to achieve a healthy body weight by dieting alone fail.[15]

Another manifestation of diet failures is the extent to which people are turning to liposuction to remove body fat. Resorting to this risky surgical procedure, which quite literally vacuums fat from under the skin, is a desperate last measure for those whose diets have failed. In 1998, there were some 400,000 liposuction procedures in the United States.[16]

For many of those who are overweight, achieving a healthy body weight depends on both reducing caloric intake and burning more calories through exercise. Metabolically, we are hunter-gatherers. Given our heritage, exercise may be a genetic imperative.

Restoring exercise in our daily lives will not be easy. Today's cities, designed for automobiles, are leading to a life-threatening level of exercise deprivation. Our health depends on creating neighborhoods that are conducive to walking, jogging, and bicycling.

The challenge is to redesign communities, making public transportation the centerpiece of urban transport, and augmenting it with sidewalks, jogging trails, and bikeways. This also means replacing parking lots with parks, playgrounds, and playing fields. Unless we can design a lifestyle that systematically restores exercise to our daily routines, the obesity epidemic—and the health deterioration associated with it—will continue to spread.

For additional information, see <www.earth-policy.org/Alerts/Alert11.htm>.

December 2001

Iran's Birth Rate Plummeting at Record Pace

Janet Larsen

Iran's population growth rate dropped from an all-time high of 3.2 percent in 1986 to just 1.2 percent in 2001, one of the fastest drops ever recorded. (See Figure 3–1.) In reducing its population growth to 1.2 percent, a rate only slightly higher than that of the United States, Iran has emerged as a model for other countries that want to accelerate the shift to smaller families.[1]

Historically, family planning in Iran has had its ups and downs. The nation's first family planning policy, introduced in 1967 under Shah Reza Pahlavi, aimed to accelerate economic growth and improve the status of women by reforming divorce laws, encouraging female employment, and acknowledging family planning as a human right.[2]

Unfortunately, this promising initiative was reversed in 1979 at the beginning of the decade-long Islamic Revolution led by Shiite Muslim spiritual leader Ayatollah Khomeini. During this period, family planning programs were seen as undue western influences and were dismantled. Health officials were ordered not to advocate contraception. During Iran's war with Iraq between 1980 and 1988, a large population was viewed as a comparative advantage, and Khomeini pushed procreation to bolster the ranks of "soldiers for Islam," aiming for "an army of 20 million."[3]

This strong pronatalist stance led to an annual population growth rate of well over 3 percent. United Nations data show Iran's population doubling from 27 million in 1968 to 55 million in 1988.[4]

During postwar reconstruction in the late 1980s, the economy faltered. Severe job shortages plagued over-crowded and polluted cities. Iran's rapid population growth was finally seen as an obstacle to development. Receptive to the nation's problems, Ayatollah Khomeini reopened dialogue on the subject of birth control. By December 1989, Iran had revived its national family planning program. Its principal goals were to encourage women to wait three to four years between pregnancies, to discourage childbearing for women younger than 18 or older than 35, and to limit family size to three children.[5]

In May of 1993, the Iranian government passed a national family planning law that encouraged couples to have fewer children by restricting maternity leave benefits after three children. It also called for the Ministries of Education, of Culture and Higher Education, and of

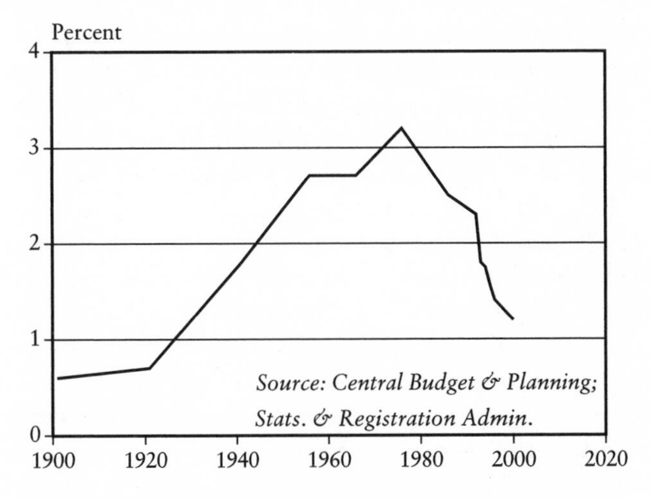

Figure 3–1. *Annual Population Increase in Iran, 1901–2000*

Health and Medical Education to incorporate information on population, family planning, and mother and child health care in curriculum materials. The Ministry of Islamic Culture and Guidance was told to allow the media to raise awareness of population issues and family planning programs, and the Islamic Republic of Iran Broadcasting was entrusted with airing such information. Money saved on reduced maternity leave funds these educational programs.[6]

From 1986 to 2001, Iran's total fertility—the average number of children born to a woman in her lifetime—plummeted from seven to less than three. The United Nations projects that by 2010 total fertility will drop to two, which is replacement-level fertility.[7]

Strong government support has facilitated Iran's demographic transition. Under the current president, Mohammad Khatami, the government covers 80 percent of family planning costs. A comprehensive health network made up of mobile clinics and 15,000 "health houses" provides family planning and health services to four fifths of Iran's rural population. Almost all of these health care centers were established after 1990. Because family planning is integrated with primary health care, there is little stigma attached to modern contraceptives.[8]

Religious leaders have become involved with the crusade for smaller families, citing them as a social responsibility in their weekly sermons. They also have issued fatwas, religious edicts with the strength of court orders, that permit and encourage the use of all types of contraception, including permanent male and female sterilization—a first among Muslim countries. Birth control, including the provision of condoms, pills, and sterilization, is free.[9]

One of the strengths of Iran's promotion of family planning is the involvement of men. Iran is the only country in the world that requires both men and women to take a class on modern contraception before receiving a marriage license. And it is the only country in the region with a government-sanctioned condom factory. In the past four years, some 220,000 Iranian men have had a vasectomy. While vasectomies still account for only 3 percent of contraception, compared with female sterilization at 28 percent, men nonetheless are assuming more responsibility for family planning.[10]

Rising literacy and a national communications infrastructure are facilitating progress in family planning. The literacy rate for adult males increased from 48 percent in 1970 to 84 percent in 2000, nearly doubling in 30 years. Female literacy climbed even faster, rising from less than 25 percent in 1970 to more than 70 percent. Meanwhile, school enrollment grew from 60 to 90 percent. And by 1996, 70 percent of rural and 93 percent of urban households had televisions, allowing family planning information to be spread widely through the media.[11]

As one of 17 countries already facing absolute water scarcity, Iran's decision to curb its rapid population growth has helped alleviate unfolding water shortages exacerbated by the severe drought of the past three years. An estimated 37 million people, more than half the population, do not have enough water.[12]

The lack of water for irrigation has helped push Iran's wheat imports to 6.5 million tons in 2001, well above the 5.8 million tons of Japan, traditionally the world's leading importer. Total grain production dropped steeply between 1998 and 2000, from 17 million to 10 million tons, largely because of the drought. The grain area harvested has decreased steadily since 1993, rapidly shrinking grain production per person.[13]

Dwindling per capita arable land and water supplies reinforce the need for population stabilization through forward-thinking family planning programs. Had the Iranian population maintained its 1986 growth rate of 3.2 percent, it would have doubled by 2008, topping 100 million instead of the projected 78 million.[14]

Because almost 40 percent of Iran's population is under the age of 15, population momentum is strong and growth in the immediate future is inevitable. To keep growth rates low, Iran needs to continue emphasizing the social value of smaller families.[15]

Among the keys to Iran's fertility transition are universal access to health care and family planning, a dramatic rise in female literacy, mandatory premarital contraceptive counseling for couples, men's participation in family planning programs, and strong support from religious leaders. While Iran's population policies and health care infrastructure are unique, its land and water scarcity are not. Other developing countries with fast-growing populations can profit by following Iran's lead in promoting population stability.

For additional information, see <www.earth-policy.org/Updates/Update4ss.htm>.

Food, Land, and Water

February 2001

Paving the Planet: Cars and Crops Competing for Land

Lester R. Brown

As the new century begins, the competition between cars and crops for cropland is intensifying. Until now, the paving over of cropland has occurred largely in industrial countries, home to four fifths of the world's 520 million automobiles. (See Figure 3–2.) But now, more and more farmland is being sacrificed in developing countries with hungry populations, calling into question the future role of the car.[1]

Millions of hectares of cropland in the industrial world have been paved over for roads and parking lots. Each U.S. car, for example, requires on average 0.07 hectares (0.18 acres) of paved land for roads and parking space. For every five cars added to the U.S. fleet, an area the size of a football field is covered with asphalt. More often than not, cropland is paved simply because the flat, well-drained soils that are well suited for farming are also ideal for building roads. Once paved, land is not easily reclaimed. As environmentalist Rupert Cutler once noted, "Asphalt is the land's last crop."[2]

The United States, with its 214 million motor vehicles, has paved 6.3 million kilometers (3.9 million miles) of roads, enough to circle the Earth at the equator 157 times. In addition to roads, cars require parking space. Imagine a parking lot for 214 million cars and trucks. If that is too difficult, try visualizing a parking lot for 1,000 cars and then imagine what 214,000 of these would look like.[3]

However we visualize it, the U.S. area devoted to roads and parking lots covers an estimated 16 million hectares (61,000 square miles), an expanse approaching the size of the 21 million hectares that U.S. farmers planted in wheat last year. But this paving of land in industrial countries is slowing as countries approach automobile saturation. In the United States, there are three vehicles for every four people. In Western Europe and Japan, there is typically one for every two people.[4]

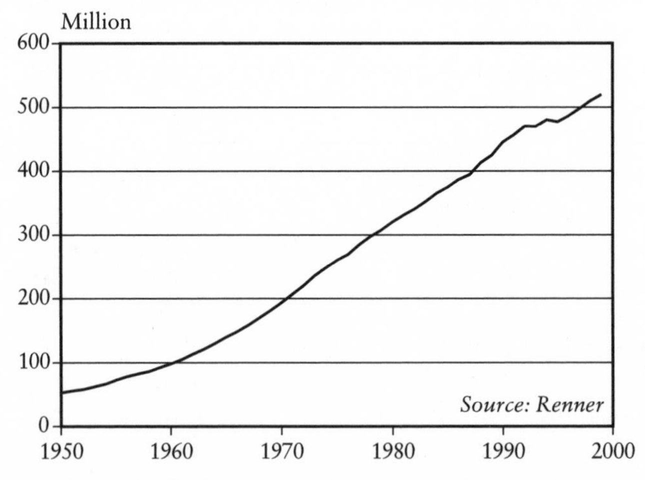

Figure 3–2. *World Automobile Fleet, 1950–99*

In developing countries, however, where automobile fleets are still small and where cropland is in short supply, the paving is just getting under way. More and more of the 11 million cars added annually to the world's vehicle fleet of 520 million are found in the developing world. This means that the war between cars and crops is being waged over wheat fields and rice paddies in countries where hunger is common. The outcome of this conflict in China and India, two countries that together contain 38 percent of the world's people, will affect food security everywhere.[5]

Car-centered industrial societies that are densely populated, such as Germany, the United Kingdom, and Japan, have paved an average of 0.02 hectares per vehicle. And they have lost some of their most productive cropland in the process. Similarly, China and India also face acute pressure on their cropland base from industrialization. Although China covers roughly the same area as the United States, its 1.3 billion people are concentrated in just one third of the country—a thousand-mile strip on the eastern and southern coast where the cropland is located.[6]

If China were one day to achieve the Japanese automobile ownership rate of one car for every two people, it would have a fleet of 640 million, compared with only 13 million today. While the idea of such an enormous fleet may seem farfetched, we need only remind ourselves that China has already overtaken the United States in steel production, fertilizer use, and red meat production. It is a huge economy and, since 1980, also the world's fastest growing economy.[7]

Assuming 0.02 hectares of paved land per vehicle in China, as in Europe and Japan, a fleet of 640 million cars would require paving nearly 13 million hectares of land, most of which would likely be cropland. This figure is over one half of China's 23 million hectares of riceland, part of which it double crops to produce 135 million tons of rice, the principal food staple. When farmers in southern China lose a hectare of double-cropped riceland to the automobile, their rice production takes a double hit. Even one car for every four people, half the Japanese ownership rate, would consume a substantial area of cropland.[8]

The situation in India is similar. While India is geographically only a third the size of China, it too has more than 1 billion people, and it now has 8 million motor vehicles. Its fast-growing villages and cities are already encroaching on its cropland. Add to this the land paved for the automobile, and India, too, will be facing a heavy loss of cropland. A country projected to add 515 million more people by 2050 cannot afford to cover valuable cropland with asphalt for roads and parking lots.[9]

There is not enough land in China, India, and other densely populated countries like Indonesia, Bangladesh, Pakistan, Iran, Egypt, and Mexico to support automobile-centered transportation systems and to feed their people. The competition between cars and crops for land is becoming a competition between the rich and the poor, between those who can afford automobiles and those who struggle to buy enough food.

Governments that subsidize an automobile infrastructure with revenues collected from the entire population are, in effect, collecting money from the poor to support the cars of the wealthy. In subsidizing the development of an auto-centered transport system, governments are also inevitably subsidizing the paving of cropland. If, as now seems likely, automobile ownership never goes beyond the affluent minority in developing countries, this becomes an ongoing and largely invisible transfer of income from the poor to the rich.

In a land-hungry world, the time has come to reassess

the future of the automobile, to design transportation systems that provide mobility for entire populations, not just affluent minorities, and that do this without threatening food security. When Beijing announced in 1994 that it planned to make the auto industry one of the growth sectors for the next few decades, a group of eminent scientists—many of them members of the Chinese Academy of Sciences—produced a white paper challenging this decision. They identified several reasons the nation should not develop a car-centered transport system, but the first was that the country did not have enough cropland both to feed its people and to provide land for the automobile.[10]

The team of scientists recommended that instead of building an automobile infrastructure of roads and parking lots, China should concentrate on developing state-of-the-art light rail systems augmented by buses and bicycles. This would not only provide mobility for far more people than a congested auto-centered system, but it would also protect cropland.[11]

There are many reasons to question the goal of building automobile-centered transportation systems everywhere, including climate change, air pollution, and traffic congestion. But the loss of cropland alone is sufficient. Nearly all of the 3 billion people to be added to the current world population of 6 billion by mid-century will be born in developing countries where there is not enough land to feed everyone and to accommodate the automobile. Future food security now depends on restructuring transportation budgets—investing less in highway infrastructure and more in rail and bicycle infrastructure.[12]

For additional information, see <www.earth-policy.org/Alerts/Alert12.htm>.

May 2001

Dust Bowl Threatening China's Future

Lester R. Brown

On April 18, scientists at the National Oceanic and Atmospheric Administration laboratory in Boulder, Colorado, reported that a huge dust storm from northern China had reached the United States, "blanketing areas from Canada to Arizona with a layer of dust." They reported that along the foothills of the Rockies, the mountains were obscured by the dust from China.[1]

This dust storm did not come as a surprise. On March 10, 2001, the *People's Daily* reported that the season's first dust storm—one of the earliest on record—had hit Beijing. These dust storms, coupled with those of last year, were among the worst in memory, signaling a widespread deterioration of the rangeland and cropland in the country's vast northwest.[2]

These huge dust plumes routinely travel hundreds of miles to populous cities in northeastern China, including Beijing, obscuring the sun, reducing visibility, slowing traffic, and closing airports. Reports of residents in eastern cities caulking windows with old rags to keep out the dust are reminiscent of the U.S. Dust Bowl of the 1930s.[3]

Eastward-moving winds often carry soil from China's northwest to North Korea, South Korea, and Japan, countries that regularly complain about dust clouds that both filter out the sunlight and cover everything with dust. Responding to pressures from their constituents, a group of 15 legislators from Japan and 8 from South

Korea are organizing a tri-national committee with Chinese lawmakers to devise a strategy to combat the dust.[4]

News reports typically attribute the dust storms to the drought of the last three years, but the drought is simply bringing a fast-deteriorating situation into focus. Human pressure on the land in northwestern China is excessive. There are too many people, too many cattle and sheep, and too many plows. Feeding 1.3 billion people, a population nearly five times that of the United States, is not an easy matter.[5]

In addition to local pressures on resources, a decision in Beijing in 1994 to require that all cropland used for construction be offset by land reclaimed elsewhere has helped create the ecological disaster that is now unfolding. In an article in *Land Use Policy*, Chinese geographers Hong Yang and Xiubein Li describe the environmental effects of this offset policy. The fast-growing coastal provinces, such as Guandong, Shandong, Xheijiang, and Jiangsu, which are losing cropland to urban expansion and industrial construction, are paying other provinces to plow new land to offset their losses. This provided an initial economic windfall for provinces in the northwest, such as Inner Mongolia (which led the way with a 22-percent cropland expansion), Gansu, Qinghai, Ningxia, and Xinjiang.[6]

As the northwestern provinces, already suffering from overplowing and overgrazing, plowed ever more marginal land, wind erosion intensified. Now accelerating wind erosion of soil and the resulting land abandonment are forcing people to migrate eastward, not unlike the U.S. westward migration from the southern Great Plains to California during the Dust Bowl years.[7]

While plows are clearing land, expanding livestock populations are denuding the land of vegetation. Following economic reforms in 1978 and the removal of controls on the size of herds and flocks that collectives could maintain, livestock populations grew rapidly. Today China has 127 million cattle compared with 98 million in the United States. Its flock of 279 million sheep and goats compares with only 9 million in the United States.[8]

In Gonge County in eastern Quinghai Province, the number of sheep that local grasslands can sustain is estimated at 3.7 million, but by the end of 1998, sheep numbers there had reached 5.5 million, far beyond the land's carrying capacity. The result is fast-deteriorating grassland, desertification, and the formation of sand dunes.[9]

In the *New York Times*, Beijing Bureau Chief Erik Eckholm writes that "the rising sands are part of a new desert forming here on the eastern edge of the Quinghai-Tibet Plateau, a legendary stretch once known for grass reaching as high as a horse's belly and home for centuries to ethnic Tibetan herders." Official estimates show 900 square miles (2,330 square kilometers) of land going to desert each year. An area several times as large is suffering a decline in productivity as it is degraded by overuse.[10]

In addition to the direct damage from overplowing and overgrazing, the northern half of China is literally drying out as rainfall declines and aquifers are depleted by overpumping. Water tables are falling almost everywhere, gradually altering the region's hydrology. As water tables fall, springs dry up, streams no longer flow, lakes disappear, and rivers run dry. U.S. satellites, which have been monitoring land use in China for some 30 years, show that literally thousands of lakes in the north have disappeared.[11]

Deforestation in southern and eastern China is reducing the moisture transported inland from the South China Sea, the East China Sea, and the Yellow Sea, writes Wang Hongchang, a Fellow at the Chinese Academy of

Social Sciences. Where land is forested, the water is held and evaporates to be carried further inland. When tree cover is removed, the initial rainfall from the inland-moving, moisture-laden air simply runs off and returns to the sea. As this recycling of rainfall inland is weakened by deforestation, rainfall in the interior is declining.[12]

Reversing this degradation means stabilizing population and planting trees everywhere possible to help recycle rainfall inland. It means converting highly erodible cropland back to grassland or woodland, reducing the livestock population, and planting tree shelter belts across the windswept areas of cropland, as U.S. farmers did to end dust storms in the 1930s.[13]

In addition, another interesting option now presents itself—the use of wind turbines as windbreaks to reduce wind speed and soil erosion. With the cost of wind-generated electricity now competitive with that generated from fossil fuels, constructing rows of wind turbines in strategic areas to slow the wind could greatly reduce the erosion of soil. This also affords an opportunity to phase out the use of wood for fuel, thus lightening the pressure on forests.[14]

The economics are extraordinarily attractive. In the U.S. Great Plains, under conditions similar to China's northwest, a large advanced-design wind turbine occupying a tenth of a hectare of land can produce $100,000 worth of electricity per year. This source of rural economic regeneration fits in nicely with China's plan to develop the impoverished northwest.[15]

Reversing desertification will require a huge effort, but if the dust bowl continues to spread, it will not only undermine the economy, but also trigger a massive migration eastward. The options are clear: Reduce livestock populations to a sustainable level or face heavy livestock losses as grassland turns to desert. Return highly erodible cropland to grassland or lose all of the land's productive capacity as it turns to desert. Construct windbreaks with a combination of trees and, where feasible, wind turbines, to slow the wind or face even more soil losses and dust storms.

If China cannot quickly arrest the trends of deterioration, the growth of the dust bowl could acquire an irreversible momentum. What is at stake is not just China's soil, but its future.

For additional information, see <www.earth-policy.org/Alerts/Alert13.htm>.

October 2001

Worsening Water Shortages Threaten China's Food Security

Lester R. Brown

A little-noticed survey released in Beijing in mid-August reveals that China's water situation is far more serious than realized. The water table under the North China Plain, which produces over half of China's wheat and a third of its corn, is falling faster than thought.[1]

Overpumping has largely depleted the shallow aquifer, reducing the amount of water that can be pumped from it to the amount of recharge from precipitation. This is forcing well drillers to go down to the region's deep aquifer, which, unfortunately, is not replenishable.[2]

The study, conducted by the Geological Environmental Monitoring Institute (GEMI) in Beijing, reported that under Heibei Province in the heart of the North China Plain, the average level of the deep aquifer dropped 2.9 meters (nearly 10 feet) in 2000. Around some cities in the province, it fell by 6 meters.[3]

He Qingcheng, head of the GEMI groundwater monitoring team, believes the fast-deteriorating water situation should be getting far more official attention. He notes that with depletion of the deep aquifer under the North China Plain, the region is losing its last water reserve—its only safety cushion.[4]

His concerns are mirrored in a new World Bank report that says, "Anecdotal evidence suggests that deep wells [drilled] around Beijing now have to reach 1,000 meters (more than half a mile) to tap fresh water, adding dramatically to the cost of supply." In unusually strong language for a Bank report, it forecasts "catastrophic consequences for future generations" unless water use and supply can quickly be brought back into balance.[5]

Further evidence of the gravity of the water situation in the North China Plain can be seen in the frenzy of well drilling in recent years. At the end of 1996, the five provinces of the North China Plain—Hebei, Henan, Shandong, and the city provinces of Beijing and Tianjin—had 2.6 million wells, the bulk of them for irrigation. During 1997, 99,900 wells were abandoned as they ran dry. Some 221,900 new wells were drilled. The desperate quest for water in China is evident as well drillers chase the water table downward.[6]

The northern half of China is drying out. Demands on the three rivers that flow eastward into the North China Plain—the Hai, the Yellow, and the Huai—are excessive, leading them to run dry during the dry season, sometimes for extended periods of time. The flow of the Yellow River into Shandong Province—the last of the eight provinces it flows through en route to the sea, and China's leading grain-producing province—has been reduced from 40 billion cubic meters (1 cubic meter = 1 ton) a year in the early 1980s to 25 billon cubic meters during the 1990s.[7]

As water tables fall, springs dry up, streams cease to flow, rivers run dry, and lakes disappear. Hebei Province once had 1,052 lakes. Only 83 remain.[8]

The water deficit in the North China Plain, the excess of use over the sustainable supply, may now exceed 40 billion tons per year. At present that deficit is being filled by groundwater mining, but when aquifers are depleted and there is nothing more to mine, the water supply will fall precipitously. In the Hai River basin—where industry and cities, including Beijing and Tianjin, now get priority—irrigated agriculture could largely disappear by 2010, forcing a shift back to less productive rain-fed agriculture.[9]

Between now and 2010, when China's population is projected to grow by 126 million, the World Bank projects that the country's urban water demand will increase from 50 billion cubic meters to 80 billion, a growth of 60 percent. Industrial water demand, meanwhile, will increase from 127 billion to 206 billion cubic meters, an expansion of 62 percent.[10]

With water worth easily 70 times as much in industry as in agriculture, farmers almost always lose in the competition with cities. As water tables continue to fall, rising pumping costs will make underground water too costly for many farmers to use for irrigation.[11]

In addition to spreading water scarcity, numerous environmental and economic forces are reducing China's grain production. As farmers attempt to maximize their income from small plots, for example, they are shifting from grain to high-value fruit and vegetable crops.[12]

China has been striving valiantly to remain self-sufficient in grain since 1994. It did so by raising support prices of grain well above the world market level, by overplowing land on a scale that helped create the world's largest dust bowl, and by overpumping the aquifers under the North China Plain.[13]

The combination of weak prices, falling water tables, and severe drought dropped the grain harvest in 2001 to 335 million tons, down from the all-time high of 392 million tons in 1998. This will fall short of projected consumption by 46 million tons. The emergence of this deficit—easily the largest in China's history—on the heels of last year's deficit of 34 million tons raises questions about future food security.[14]

The back-to-back grain shortfalls in the last two years at a time when China's imports of grain are negligible have dropped stocks by roughly 81 million tons. With its accessible stocks of grain now largely depleted, another sizable crop shortfall in 2002 would likely force China to import large amounts of grain to avoid rising food prices.[15]

China's grain imports could climb quickly, as its recent experience with soybeans shows. When grain support prices were raised in 1994, resources were diverted from soybeans—the nation's fourth ranking crop after wheat, rice, and corn. As a result, the soybean harvest has fallen 6 percent since 1994 while demand has doubled. In an abrupt turnaround, China has gone from being a small net exporter of soybeans in 1993 to being the world's largest importer in 2001, bringing in 14 million of the 30 million tons it consumes.[16]

If China has another sizable grain harvest shortfall in 2002, it will likely be forced to import grain far in excess of the 7 million tons of wheat and 5 million tons of corn that it must promise to import if it joins the World Trade Organization in late 2001, as expected.[17]

With its aquifers being depleted, China is now reconsidering its options for reestablishing a balance between water use and supply. Three possible initiatives stand out: water conservation, diversion of water from the south to the north, and grain imports. A south/north diversion to transport water from the Yangtze River basin will cost tens of billions of dollars and displace hundreds of thousands of people. A comparable investment in more water-efficient industrial practices, more water-efficient household appliances, and, above all, the use of more-efficient irrigation practices would likely yield more water. Since it takes 1,000 tons of water to produce 1 ton of grain, importing grain is the most efficient way to import water.[18]

Regardless of whether it concentrates solely on conservation or also does a south/north diversion, China will almost certainly have to turn to the world market for grain imports. If it imports even 10 percent of its grain supply—40 million tons—it will become overnight the largest grain importer, putting intense pressure on exportable grain supplies and driving up world prices. If this happens, we probably won't need to read about it in the newspapers. It will be evident at the supermarket checkout counter.

For additional information, see <www.earth-policy.org/Updates/Update1.htm>.

February 2002

World's Rangelands Deteriorating Under Mounting Pressure

Lester R. Brown

In late January, a dust storm originating in northwestern China engulfed Lhasa, the capital of Tibet, closing the airport for three days and disrupting tourism. Such dust storms are no longer uncommon. Dust storms originating in Central Asia, coupled with those originating in Saharan Africa that now frequently reach the Caribbean, remind us that desertification of the world's rangelands is ongoing.[1]

Even though the damage from overgrazing is spreading, the world's livestock population continues to grow, tracking the growth in human population. As world population increased from 2.5 billion in 1950 to 6.1 billion in 2001, the world's cattle herd went from 720 million to 1.53 billion. The number of sheep and goats expanded from 1.04 billion to 1.75 billion.[2]

With 180 million pastoralists worldwide now trying to make a living tending 3.3 billion cattle, sheep, and goats, grasslands are under heavy pressure. As a result of overstocking, grasslands are deteriorating in much of Africa, the Middle East, Central Asia, the northern part of the Indian subcontinent, Mongolia, and much of northern China. Overgrazing of rangelands initially reduces their productivity but eventually it destroys them, leaving desert. Degraded rangeland, worldwide, totals 680 million hectares—five times the U.S. cropland area.[3]

Rangelands, consisting almost entirely of land that is too dry or too steeply sloping to support crop production, account for one fifth of the earth's land surface, more than double the area that is cropped. Tapping the productivity of this vast area depends on ruminants—cattle, sheep, and goats—animals whose complex digestive systems enable them to convert roughage into food, including beef, mutton, and milk, and industrial materials, importantly leather and wool.[4]

Some four fifths of world beef and mutton production, roughly 52 million tons, comes from animals that forage on rangelands. In Africa, where grain is scarce, 230 million cattle, 246 million sheep, and 175 million goats are supported almost entirely by grazing and browsing. The number of livestock, a cornerstone of many African economies, often exceeds grassland carrying capacity by half or more. A study that charted the mounting pressures on grasslands in nine southern African countries found that the capacity of the land to sustain livestock is diminishing.[5]

Fodder needs of livestock in nearly all developing countries now exceed the sustainable yield of rangelands and other forage resources. In India, with the world's largest cattle herd, the demand for fodder in 2000 was projected at 700 million tons, while the sustainable supply totaled just 540 million tons. A report from New Delhi indicates that in states with the most serious land degradation, such as Rajasthan and Karnataka, fodder supplies satisfy only 50–80 percent of needs, leaving large numbers of emaciated, unproductive cattle.[6]

China faces similarly difficult challenges. The northwest of China, where there are no landownership rights and no fences, has become a vast grazing commons. Since the economic reforms of 1978, there has been little incentive for individual families to limit the size of their flocks and herds. As a result, livestock numbers have soared. The United States, which has a comparable grazing capacity, has 98 million head of cattle while China has 130 million head. But the big difference is in the number of sheep and goats: 9 million in the United States, 290 million in China.[7]

In Gonge County, for example, in eastern Qinghai Province, the local grasslands can support an estimated 3.7 million sheep. But by the end of 1998, the region's flock had reached 5.5 million—far beyond its carrying capacity. The result is fast-deteriorating grassland and the creation of a new desert, replete with sand dunes.[8]

The mounting pressures on rangelands in the Middle East are illustrated by Iran, a country of 71 million people. The 8 million cattle and 81 million sheep and goats that graze its rangelands supply not only milk and meat, but also the wool for the country's fabled rug-making industry. In a land where sheep and goats outnumber humans, and where rangelands are being pushed to their limits, the current livestock population may not be sustainable.[9]

Land degradation from overgrazing is taking a heavy economic toll in lost livestock productivity. In the early stages of overgrazing, the costs show up as lower land productivity. But if the process continues, it destroys vegetation, leading to the erosion of soil and the eventual creation of wasteland. A U.N. assessment of the earth's dryland regions, done in 1991, estimated that livestock production losses from rangeland degradation exceeded $23 billion. (See Table 3–3.)[10]

In Africa, the annual loss of rangeland productivity is estimated at $7 billion, more than the gross domestic product of Ethiopia. In Asia, livestock losses from rangeland degradation total over $8 billion. Together, Africa and Asia account for two thirds of the global loss.[11]

Table 3–3. *Economic Cost of Rangeland Degradation*

Region	Average Annual Income Forgone
	(million U.S. dollars)
Africa	6,966
Asia	8,313
Australia	2,529
Europe	564
North America	2,878
South America	2,084
Total	23,334

Source: H. Dregne et al., "A New Assessment of the World Status of Desertification," *Desertification Control Bulletin*, no. 20, 1991.

Arresting the deterioration of the world's rangelands presents a difficult challenge. One key to slowing the growth in livestock populations is to stop the growth in human numbers. Iran, recognizing the threat of overgrazing and other population-related stresses it was facing some 15 years ago, dropped its population growth from 4 percent a year to scarcely 1 percent in 2001, illustrating what can be done with committed leadership.[12]

Another key to lightening pressure on rangelands is the spreading practice of feeding livestock crop residues that would otherwise be burned, either because they are needed for fuel or because double cropping requires destruction of the residues. India has been uniquely successful in converting crop residues into milk—expanding production from 20 million tons in 1961 to 80 million tons in 2001, and without feeding grain. Its farmers did so almost entirely by using crop residues and by stall-feeding grass cut and collected by hand.[13]

China also has a large potential to feed corn stalks and wheat and rice straw to cattle or sheep. As the world's leading producer of both rice and wheat and the second-ranked producer of corn, China annually harvests an estimated 500 million tons of straw, corn stalks, and other crop residues. Feeding crop residues in the major crop-producing provinces of east central China—Hebei, Shandong, Henan, and Anhui—has created a "Beef Belt," whose beef output dwarfs that of the northwestern grazing provinces of Inner Mongolia, Qinghai, and Xinjiang.[14]

In rangeland reclamation, where successes are few, a promising low-cost technique for reclaiming overgrazed and exhausted rangeland is being developed at the International Center for Agricultural Research in the Dry Areas (ICARDA) in Syria. ICARDA scientists have developed a simple implement that slightly depresses the soil in double rows 20 centimeters (8 inches) apart. The implement seeds grass in these twin depressions, which follow the contour of the land, enabling them to trap rainwater runoff and restore vegetation.[15]

It will take an enormous effort to stabilize livestock populations at a sustainable level and to restore the world's degraded rangelands. This will be costly, but failing to halt the desertification of rangelands will be even costlier as flocks and herds eventually shrink and as the resulting poverty forces large-scale migration from the affected areas.

For additional information, see <www.earth-policy.org/Updates/Update6.htm>.

Fisheries, Forests, and Disappearing Species

October 2000

Fish Farming May Overtake Cattle Ranching as a Food Source

Lester R. Brown

Aquacultural output, growing at 11 percent a year over the past decade, is the fastest-growing sector of the world food economy. Climbing from 13 million tons of fish produced in 1990 to 31 million tons in 1998, fish farming is poised to overtake cattle ranching as a food source by the end of this decade.[1]

This record aquacultural growth is signaling a basic shift in our diet. Over the last century, the world relied heavily on two natural systems—oceanic fisheries and rangelands—to satisfy a growing demand for animal protein, but that era is ending as both systems are reaching their productive limits. Between 1950 and 1990, beef production, four fifths of it from rangelands, nearly tripled, climbing from 19 million to 53 million tons before plateauing. (See Figure 3–3.) Meanwhile, the oceanic fish catch grew from 19 million to 86 million tons, more than quadrupling, before leveling off. Since 1990, there has been little growth in either beef production or the oceanic fish catch.[2]

Additional production of beef or seafood now depends on placing more cattle in feedlots or more fish in ponds. At this point, the efficiency with which cattle and fish convert grain into protein begins to reshape production trends and thus our diets. Cattle require some 7 kilograms of grain to add 1 kilogram of live weight, whereas fish can add a kilogram of live weight with less than 2 kilograms of grain. Water scarcity is also a matter of concern since it takes 1,000 tons of water to produce 1 ton of grain. But the fish farming advantage in the efficiency of grain conversion translates into a comparable advantage in water efficiency as well, even when the relatively small amount of water for fish ponds is included. In a world of land and water scarcity, the advantage of fish ponds over feedlots in producing low-cost animal protein is clear.[3]

In contrast to meat production, which is concentrated in industrial countries, some 85 percent of fish farming is in developing countries. China, where fish farming began

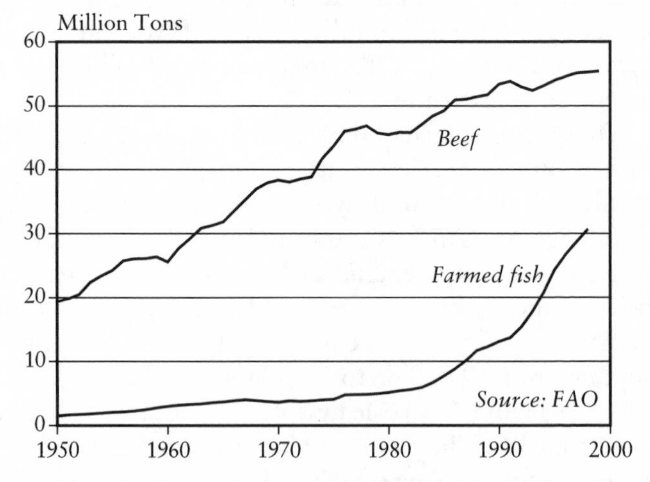

Figure 3–3. *World Aquacultural Production, 1950–98, and Beef Production, 1950–99*

more than 3,000 years ago, accounted for 21 million tons of the 31 million tons of world aquacultural output in 1998. India is a distant second with 2 million tons. Other developing countries with thriving aquacultural sectors include Bangladesh, Indonesia, and Thailand.[4]

Among industrial countries, Japan, the United States, and Norway are the leaders. Japan's output of 800,000 tons consists of high-value species, such as scallops, oysters, and yellowtail. The U.S. output of 450,000 tons is mostly catfish. Norway's 400,000 tons is mostly salmon.[5]

With overfishing now commonplace, developing countries are turning to fish farming to satisfy their growing appetite for seafood largely because the oceanic option is not available to them as it was earlier to industrial countries. For example, as population pressure on the land intensified in Japan over time, it turned to the oceans for its animal protein, using scarce land for rice. Today Japan's 125 million people consume some 10 million tons of seafood each year. If China's 1.25 billion were to eat seafood at the same rate, they would need 100 million tons—the global fish catch.[6]

Although at least 220 species of fin fish, shellfish, and crustaceans are farmed commercially, a dozen or so dominate world output. Among the fin fish, five species of carp—all widely grown in China—lead the way with a combined output of some 11 million tons in 1998, more than a third of world aquacultural output. Among shellfish, the Pacific cupped oyster, at 3.4 million tons (including shell), dominates, followed by the Yesso scallop and the blue mussel.[7]

In China, fish are produced primarily in ponds, lakes, reservoirs, and rice paddies. Some 5 million hectares of land are devoted exclusively to fish farming, much of it in carp polyculture. In addition, 1.7 million hectares of riceland are used to produce rice and fish together.[8]

Over time, China has evolved a fish polyculture using four types of carp that feed at different levels of the food chain. Silver carp and bighead carp are filter feeders, feeding on phytoplankton and zooplankton, respectively. The grass carp, as its name implies, feeds largely on vegetation, while the common carp is a bottom feeder, living on detritus that settles to the bottom. Most of China's aquaculture is integrated with agriculture, enabling farmers to use agricultural wastes, such as pig manure, to fertilize ponds, thus stimulating the growth of plankton. Fish polyculture, which typically boosts the fish yield per hectare over that of monocultures by at least half, also dominates fish farming in India.[9]

As land and water become scarce, China's fish farmers are intensifying production by feeding more grain concentrates to raise pond productivity. Between 1990 and 1996, China's farmers raised the annual pond yield per hectare from 2.4 tons of fish to 4.1 tons.[10]

In the United States, catfish, which require only 1.6 kilograms of feed to gain 1 kilogram of live weight, is the leading aquacultural product. With U.S. catfish production last year at roughly 600 million pounds (270,000 tons), or more than 2 pounds for each American, U.S. consumption of catfish exceeded that of lamb and mutton. Catfish production is concentrated in four states: Mississippi, Louisiana, Alabama, and Arkansas. Mississippi, with some 174 square miles (45,000 hectares) of catfish ponds and easily 60 percent of U.S. output, is the catfish capital of the world.[11]

Among the aquatic species that are widely farmed, two especially wreak extensive environmental havoc—salmon, with production of 700,000 tons per year, and shrimp at 1.1 million tons per year. Salmon are grown mostly in industrial countries, principally in Norway, for consumption in those countries. Shrimp, by contrast, are grown largely in developing countries, importantly Thailand, Ecuador, and Indonesia, for export to more affluent societies.[12]

Salmon, a carnivorous fish, are fed a diet consisting primarily of fishmeal that is typically made from anchovies, herring, or the remnants of fish processing. In stark contrast to the production of herbivorous species, such as carp and catfish, which lighten the pressure on oceanic fisheries, salmon production actually intensifies pressure because it requires up to 5 tons of landed fish for each ton of salmon produced.[13]

Another concern is that if farmed salmon, which are bred for fast growth and not for survival in the wild, escape because of damage to the pens by storms or attacks by predators, such as harbor seals, they can breed with wild salmon, weakening the latter's capacity to survive. Fish grown in offshore cages or pens, as salmon frequently are, also concentrate large quantities of waste, which itself presents a management problem. For example, the waste produced by farmed salmon in Norway is roughly equal to the sewage produced by Norway's 4 million people.[14]

Shrimp are often produced by clearing coastal mangrove forests, which protect coastlines and serve as nurseries for local fish. Mangrove destruction can cause a decline of local fisheries that will actually exceed the gains from shrimp production, leading to a net protein loss. In addition, because shrimp rations are also high in fishmeal, shrimp, like salmon, put additional pressure on oceanic fisheries.[15]

A world that is reaching the limits with both oceanic fisheries and rangelands while adding 80 million people each year needs efficient new sources of animal protein. Herbivorous species of fish, such as carp grown in polycultures, carp grown in combination with rice, or catfish grown in ponds, offer a highly efficient way of expanding animal protein supplies in a protein-hungry world. Fish farming is not a solution to the world food problem, but as China has demonstrated, it does offer a potential source of low-cost animal protein for lower-income populations. The forces that have made aquaculture the world's fastest-growing source of animal protein over the last decade are likely to make it the fastest-growing source during this decade as well.[16]

For additional information, see <www.earth-policy.org/Alerts/Alert9.htm>.

March 2002

Our Closest Relatives Are Disappearing

Janet Larsen

After more than a century of no known primate extinctions, scientists recently confirmed the disappearance of a subspecies of a West African monkey. The loss of this monkey, known as Miss Waldron's red colobus, may be a harbinger of future losses of our closest evolutionary relatives.[1]

Out of some 240 known primate species, 19 are critically endangered, up from 13 in 1996. This classification refers to species that have suffered extreme and rapid reductions in population or habitat. Their remaining numbers range from less than a few hundred to, at most, a few thousand individuals. If their populations continue to shrink at recent rates, some species will not survive this decade. This group, according to the World Conservation Union–IUCN's *2000 Red List of Threatened Species*, includes eight monkeys from Brazil's Atlantic rainforest, where 97 percent of the forest has been lost, two apes and a monkey from Indonesia, three monkeys from Viet Nam, one each from Kenya and Peru, and three lemur species from Madagascar.[2]

At the endangered level, IUCN's next degree of threat, there are 46 primate species, up from 29 in 1996. These species face a very high probability of extinction, some within the next 20 years. An additional 51 species are listed as vulnerable. These primates have slightly larger populations but still may disappear within this century.

Critically endangered, endangered, and vulnerable species together total 116, or nearly half of the 240 or so primate species.[3]

When the last Ice Age ended 10,000 years ago, baboons outnumbered humans by at least two to one. If all non-human primate populations were counted together, including the large populations of some of the smaller species, they dwarfed the human population. Now that has changed. The development of agriculture allowed for rapid human population growth, and about 2,000 years ago, humans—numbering 300 million—became the most abundant of the primates. By 1930, the human population of 2 billion likely outnumbered all other primates combined.[4]

Today, at 6.1 billion and climbing, we are threatening the survival of many of our primate cousins, including our closest living relatives, the chimpanzees and bonobos, with which we share over 98 percent of our genome. The other apes are quite close to us as well, not only genetically, but also in observed behavior. Yet with the 300,000 human babies born each day exceeding the total population of the great apes, even our evolutionary proximity may not prevent us from eradicating our near-kin.[5]

While humans now inhabit every corner of the earth, most other primates exhibit strong endemism, meaning that a species is restricted to a particular area. Almost three quarters of all primates live in just four countries: Brazil, the Democratic Republic of the Congo (formerly Zaire), Indonesia, and Madagascar. In each of these countries, forest cover is decreasing. Because habitat loss is a danger to 90 percent of threatened primates, their concentration in a few countries greatly increases their vulnerability.[6]

In Indonesia, diverse forests and wild inhabitants have suffered from logging fueled by corruption and political instability. Within the past decade, deforestation rates doubled, claiming almost 2 million hectares each year. As deforestation rates doubled, orangutan numbers dropped by half. By 2005, the country faces the loss of all lowland forest from Sumatra, and thus the extinction of the critically endangered Sumatran orangutan, among many other species. The Borneo orangutan, after suffering from logging, hunting, and the catastrophic fires of 1997, is not likely to survive beyond 2010 if current trends continue.[7]

Our closest relative, the bonobo, is endemic to the Congo, a country plagued by civil war and occupation by foreign military and rebel groups. Along with many other primates in the region, the slow-breeding bonobo has seen a rapid decline. In 1980 there were close to 100,000 bonobos; now there may be fewer than 10,000.[8]

Although the civil war has created millions of human refugees and may have elevated the demand for meat from wild animals (bushmeat), the resulting sluggish economic development may have slowed logging in the Congo, the country containing half of Africa's remaining tropical moist forests. If political stability returns, tree cutting could increase severalfold in the next few years, accelerating what could be the first great ape extinction.[9]

Gorilla populations have dropped to dangerously low levels, largely from illegal commercial bushmeat hunting. Fewer than 325 mountain gorillas exist, and all are in one subpopulation spanning Rwanda, the Congo, and Uganda. The rarest, the Cross River Gorilla, is limited to only 150–200 individuals scattered among several lingering subpopulations on the Cameroon/Nigeria border region.[10]

In parts of West and Central Africa, hunting is an even greater threat than forest loss. There the bushmeat trade, consisting primarily of forest antelope, pigs, and primates, is worth over $1 billion a year. In areas where social turmoil has ravaged traditional economic activities, and the average annual family income is less than $100, the lure of earning $300 to $1,000 each year as a hunter has enticed many. Logging and, to a lesser extent, mining companies have penetrated forests, with their settlements increasing bushmeat demand, while their roads facilitate hunting.[11]

Exploitative hunting is not profitable in the long term, however, because wild populations, especially those of the large and slow-reproducing apes, are soon decimated. Over 1 million tons of wild meat is consumed annually in the Congo Basin, almost six times more than the forests' sustainable yield. Commercial hunting has emptied forests that were once full of animals.[12]

Though rural communities have long subsisted on wild animals and other forest foods, with up to 60 percent of their protein coming from bushmeat, most bushmeat from this region is now consumed in cities. Almost half of the 30 million people living in the forested regions of Central Africa are city-dwellers who are being fed with bushmeat from collapsing wildlife populations. As cities grow and bushmeat hunting accelerates to meet rising demand, it is estimated that hunting could eliminate all viable African ape populations in less than 20 years.[13]

To save other primates from being lost in what is considered the earth's sixth major extinction event, resources are needed to curb illegal logging and hunting. Illegal logging has ruined vast stretches of original primate habitat. Much of the bushmeat hunted comes from protected areas, and international trade in primates is already unlawful under the Convention on International Trade in Endangered Species of Wild Fauna and Flora. But when enforcement is lacking, illegal practices continue.[14]

Large wilderness blocks of biologically rich areas can

be converted to new parks that take into account the needs of wildlife and human populations. Ecotourism endeavors can be used to support primate conservation, and hunters can find alternative income in park protection work once they realize that live animals can be much more valuable than dead ones.

Understanding ourselves better—our biology, psychology, and sociology—depends in part on understanding our closest living relatives better. If we destroy them, we may never fully understand ourselves.

For additional information, see <www.earth-policy.org/Updates/Update7.htm>.

May 2002

Illegal Logging Threatens Ecological and Economic Stability

Janet Larsen

Extensive floods in Indonesia during early 2002 have killed hundreds of people, destroyed thousands of homes, damaged thousands of hectares of rice paddy fields, and inundated Indonesian insurance companies with flood-related claims. Rampant deforestation, much of it from illegal logging, has destroyed forests that stabilize soils and regulate river flow, causing record floods and landslides.[1]

In just 50 years, Indonesia's total forest cover fell from 162 million hectares to 98 million. Roads and development fragment over half of the remaining forests. More than 16 million people depend on fresh water from Indonesia's 15 largest watersheds, which between 1985 and 1997 lost at least 20 percent of their forest cover. Loggers have cleared almost all the biologically diverse lowland tropical forests off Sulawesi, and if current trends continue, such forests will be gone from Sumatra in 2005 and from Kalimantan by 2010.[2]

Domestic wood supply in Indonesia was documented at 20 million cubic meters in 2000, while demand stood at some 60 million cubic meters. Thus legal supplies of wood fiber fall short of demand by up to 40 million cubic meters per year. Illegal logging fills the gap—accounting for almost 70 percent of wood supply. All told, illegal logging alone has destroyed 10 million hectares of Indonesia's rich forests, an area the size of Virginia in the United States.[3]

Indonesia's situation is not unique. The Philippines once held 16 million hectares of forests but is now down to less than 700,000 hectares. In this country where illegal logging runs rampant, forest loss from tree felling and conversion to agriculture is cited as the cause of flooding, acute water shortages, rapid soil erosion, river siltation, and mudslides that have taken lives, destroyed properties, and wreaked environmental damage.[4]

In 1989, Thailand banned the logging of natural forests in direct response to devastating floods and landslides that had taken 400 lives the year before. Though illegal logging is now at lower levels than before the ban, it is still widespread. More recently, massive flooding of China's Yangtze River in 1998, which was linked to the removal of 85 percent of the upper river basin's original tree cover, propelled China to issue a ban on logging in the upper reaches of the Yangtze and Yellow Rivers and to begin a reforestation campaign.[5]

China consumes nearly 280 million cubic meters of timber a year, but domestic supply currently provides only 142 million cubic meters. As production shrinks, China is turning to imports and illegal logging to make up for the shortfall. The International Tropical Timber Organization forecasts that within the next few years China will become the world's largest log importer, edging out the United States and eclipsing Japan, whose massive imports have already destroyed many of the rainforests of the Philippines and much of Borneo.[6]

Fifty-seven percent of the logs brought into China originate in Russia, where poor law enforcement, corruption, and the abandonment of local timber-processing plants have led people to illegally cut trees for companies that send raw materials to China for processing. At least one fifth of Russia's timber harvest is taken illegally or drastically violates existing legislation.[7]

To China's south, Myanmar (formerly Burma) holds about half of mainland Southeast Asia's forests. These contain a variety of tropical hardwood species that are increasingly drawing interest from China. On paper, Myanmar supplies less than 10 percent of China's log imports, but industry workers say the numbers must be at least twice as high. Burmese log exports to China are growing much faster than the trees, many of which are hundreds of years old, can be replaced. In 1949, tropical forests covered 21 percent of the country's land area, but now less than 7 percent of Myanmar is forested.[8]

In Laos, where the volume of illegal logging is the equivalent of at least one sixth of the legal harvest, the army openly cuts forests. Now less than 40 percent of the country is forested, down from 70 percent in 1940. In Cambodia, over 70 percent of the timber export volume consists of unreported logs. And Viet Nam could lose all substantial forest cover by 2020 if the current rate of deforestation continues.[9]

As the growing Asian timber market has exhausted supplies over much of the continent, wood imports to Asia from Africa have steadily increased. From 1993 to 1999, Europe imported 40 percent of central African logs, but since 1996, rising demand from Asia has made that region the number one importer of African timber.[10]

Forest products are the second largest export for both Cameroon and Gabon, generating about 20 percent and 13 percent of respective export revenues. Between 1990 and 1995, the share of Cameroonian logs going to Asia soared from 7 percent to 50 percent. Unfortunately, only half the logging companies in Cameroon are licensed, and among these companies, violations such as felling trees smaller than the legal size and cutting outside concession boundaries are common.[11]

These examples cover only a portion of the global timber market. Uncontrolled deforestation abounds in other countries—in Brazil, with the world's highest deforestation rate, where an estimated 80 percent of logging is illegal; in Mexico, which is losing over 1 million hectares each year; and in Ethiopia, where in just 40 years forest cover has plummeted from around 40 million hectares to 2.7 million, only half of which is natural forest. Rarely, though, is deforestation purely a local issue.[12]

The world's eight largest industrial countries plus the rest of the European Union buy 280 million cubic meters of timber and timber products from abroad each year, accounting for 74 percent of the world's timber imports. Most of this wood comes from countries where illegal tree felling is the norm. In 2000, the United States alone imported over $450 million worth of timber from Indonesia, which given Indonesia's illegal logging rate could represent $330 million worth of timber from illegal sources.[13]

If importing countries insist that timber and timber products are certified under internationally recognized environmental and social standards like those of the Forest Stewardship Council, illegal logging becomes more difficult. Exporting countries would profit by protecting the integrity of forest ecosystems, and could secure higher prices for certified wood on international markets. Russia, for instance, which loses $1 billion in export revenues each year because its wood is not certified, is now developing a mandatory certification system for standing forests.[14]

Certification along with existing international agreements, such as the Convention on International Trade in Endangered Species of Wild Fauna and Flora, can help to prevent illegal logs from crossing international borders—if laws and standards are upheld. Recycling and reduced use of throwaway timber products can lower the demand for timber that has made illegal logging profitable. As the Chinese government has recognized, the services that forests provide, such as flood control, can be worth far more than the lumber they contain.

For additional information, see <www.earth-policy.org/Updates/Update11.htm>.

Ecological Economics

April 2002

Green Power Purchases Growing

Bernie Fischlowitz-Roberts

In June 2001, the city of Chicago and 48 city government agencies signed a contract with local utility ComEd to purchase 10 percent of their electricity from renewable sources, a figure due to increase to 20 percent in five years. This is the largest such purchase in the United States, but Chicago is just one example of the many cities, businesses, and individuals who are buying "green power." Utilities in eight states and many other industrial countries now offer such purchases.[1]

In October 1999, Leeds Metropolitan University in the United Kingdom started buying at least 30 percent of its energy from green power. Six months later, Edinburgh University signed an agreement to obtain 40 percent of its energy this way. Since renewable energy sources in the United Kingdom are exempt from a climate change levy enacted in April 2001, making this switch is virtually cost-free and can even save money.[2]

The Netherlands has more than 775,000 green energy customers, which represents 5 percent of the population.

The number of customers has tripled in just one year. This rapid growth is due to an energy tax exemption for green electricity, green energy deregulation, and successful marketing campaigns. With Dutch demand outstripping supply, more than 30 percent of the green power used there is now imported.[3]

Germany has approximately 280,000 green energy customers. Many large German companies are buying green power, helping to create consumer demand to move beyond fossil fuels. Dresdner Bank, Weleda AG, a large homeopathic medicine company, and 23 kindergartens in Lorrach all purchase 100 percent green power.[4]

In March 1999, a comprehensive ecological tax reform law took effect in Germany that reduced income taxes, raised taxes on energy sources tied to carbon emissions, and exempted renewables. In February 2000, the parliament passed a renewable energies law that included payments for excess green energy generation fed back into the power grid; at those times, the meters run backwards, reducing customers' electric bills. These policies, which help make green energy cost-effective, are essential to the ultimate success of green power programs.[5]

Australia's green power sales are evenly divided between 60,000 residential customers and almost 2,500 commercial ones. Most of the green energy supplied to date in Australia is derived from biomass and hydroelectric power, with only 8 percent coming from wind or solar. With wind resource development accelerating, however, wind's share is increasing rapidly.[6]

In Colorado in the United States, the Grassroots Campaign for Wind Power has educated citizens about the benefits of wind power and encouraged a shift in purchasing behavior. As a result, Colorado has 20,000 residential green power subscribers and numerous commercial ones, including IBM, Hewlett-Packard, and

Patagonia, as well as the cities of Denver, Fort Collins, and Aspen. Even the governor's mansion buys green power. At the University of Colorado, students voted overwhelmingly during the spring of 2000 to raise student fees by $1 per semester in order to purchase wind power. This fee increase generates $50,000 per year, enough to buy the output of one wind turbine, or 2 million kilowatt-hours of electricity.[7]

A large number of U.S. businesses and other commercial customers have also signed up. In addition to large, high-profile companies like Toyota and Kinko's, lesser-known companies are aligning their purchasing decisions with their environmental values. Fetzer Vineyards, for example, began buying 5 million kilowatt-hours of renewable energy annually for its organic wine operations in Hopland, California.[8]

In 1996, Salem, Oregon, was the first U.S. city to go completely renewable for all power used in the city. Already getting 83 percent of its electricity from hydropower, it replaced the remaining 17 percent, which was from fossil fuels and nuclear power, with wind energy purchased from the Bonneville Power Administration (BPA). In 2000, Oakland, California, signed up for 9 megawatts of green power to meet its entire electricity load for city agencies. Santa Monica, California, also uses exclusively green power for its city facilities.[9]

Government agencies are also signing up for green power. The U.S. Environmental Protection Agency (EPA) purchases 100 percent green power at five of its facilities across the country. In so doing, EPA currently obtains 9 percent of its overall electricity consumption from green power. In 2000, Secretary of Energy Bill Richardson directed the U.S. Department of Energy to purchase 3 percent of total electricity needs from non-hydro renewable sources by 2005, and 7.5 percent by 2010.[10]

Green power offers an opportunity for citizens and corporations to act on their environmental concerns and to demonstrate support for public policies supporting renewable energy. In Colorado, for example, the demand for green power is driving the investment in wind farms, a fast-growing source of power in the state.

It is clear, however, that green power purchase options alone, even in fully deregulated markets, will not bring about the large-scale changes needed to move the world to a sustainable energy economy. Individual and corporate choices based on environmental concerns cannot replace the role of public policies. Indeed, tax restructuring and renewables portfolio standards, acting in concert with energy efficiency and green power programs, represent the best hope for creating an ecologically sustainable energy economy.

To be certified as a "green-e" product in the United States by the Center for Resource Solutions (CRS), a voluntary program, green power offerings must contain more than half renewable energy. Thus in many cases, almost half of the mix can come from fossil fuels and nuclear power. CRS set up the 50-percent standard mindful of the need for wide acceptance by various stakeholders, and wary of setting the initial standard too high for many companies to meet it. While such concerns are important, the ideal green power products would emphasize wind, solar, and geothermal, since they do not contribute to climate change, air pollution, or acid rain. Fossil fuels and nuclear power would be excluded from such products.[11]

The new green power standard in Illinois, unveiled by environmental and consumer groups in the state, is the greenest in the United States. To qualify, green power in Illinois must be from new renewable sources, must be composed of at least two thirds wind and solar power, and must create air quality benefits for the state. A similar standard, if adopted nationwide, could yield substantial benefits.[12]

The green power option for consumers and businesses is generating demand, yet its current definitions are flawed. Unless standards require much higher percentages of renewables that are green, customers may be paying a premium for only marginally cleaner power. To address climate change, the global energy economy needs to be fundamentally restructured. Green power purchase options, one instrument among many to do this, can help move us in the right direction.

For additional information, see <www.earth-policy.org/Updates/Update9.htm>.

April 2002

New York: Garbage Capital of the World

Lester R. Brown

The question of what to do with the 11,000 tons of garbage produced each day in New York City has again surfaced, this time with Mayor Michael Bloomberg's budget, which proposes to halt the recycling of metal, glass, and plastic to save money. Unfortunately, this would mean more garbage to dispose of when the goal should be less.[1]

The city's garbage problem has three faces. It is an economic problem, an environmental challenge, and a potential public relations nightmare. When the Fresh Kills landfill, the local destination for New York's garbage, was permanently closed in March 2001, the city found itself hauling garbage to distant landfill sites in New Jersey, Pennsylvania, and Virginia—some of them 300 miles away.[2]

Assuming a load of 20 tons of garbage for each of the tractor trailers used for the long-distance hauling, some 550 rigs are needed to move garbage from New York City each day. These tractor trailers form a convoy nearly nine miles long, impeding traffic, polluting the air, and raising carbon emissions. This daily convoy led Deputy Mayor Joseph J. Lhota, who supervised the Fresh Kills shutdown, to say that getting rid of the city's trash is now "like a military-style operation on a daily basis."[3]

Instead of rapidly reducing the amount of garbage generated as Fresh Kills was filling, the decision was made simply to haul it all elsewhere. Fiscally strapped local communities in other states are willing to take New York's garbage—if they are paid enough. Some see it as a bonanza. For the state governments, however, that are saddled with increased road maintenance costs, the arrangement is not so attractive. They also have to contend with the traffic congestion, noise, increased air pollution, and complaints from nearby communities.

Virginia Governor Jim Gilmore wrote to Mayor Rudy Giuliani in 2001 complaining about the use of Virginia as a dumping ground. "I understand the problem New York faces," he noted. "But the home state of Washington, Jefferson and Madison has no intention of becoming New York's dumping ground."[4]

The new governor of Virginia, Mark Warner, proposed in early April 2002 a tax of $5 per ton on all solid waste deposited in Virginia. This is expected to generate an annual cash flow of $76 million for the Virginia treasury, but it will not help New York with its economic woes.[5]

In Pennsylvania, the General Assembly is considering legislation that would restrict garbage imports from other states. As landfills in adjacent states begin to fill up, there will be progressively fewer sites to take New York's garbage, pushing disposal costs ever higher.[6]

Landfilling garbage uses land. For every 40,000 tons of garbage added to a landfill, at least one acre of land is lost to future use. A large surrounding area is also lost, as the landfill with its potentially toxic wastes must be isolated from residential areas.[7]

Mayor Bloomberg's office has proposed incineration as the solution to the garbage mess. But burning 11,000 tons of garbage each day will only add to air pollution, making already unhealthy city air even worse. Like hauling the garbage to distant sites, incineration treats the symptoms, not the causes of New York's mountain of garbage.[8]

The amount of garbage produced in the city is a manifestation of a more fundamental problem—the evolution of a global throwaway economy. Throwaway products, facilitated by the appeal to convenience and the artificially low cost of energy, account for much of the garbage we produce.

It is easy to forget how many throwaway products there are until we actually begin making a list. We have substituted facial tissues for handkerchiefs, disposable paper towels for hand towels, disposable table napkins for cloth napkins, and throwaway beverage containers for refillable ones. In perhaps the ultimate insult, the shopping bags that are used to carry home throwaway products are themselves designed to be discarded, becoming part of the garbage flow. The question at the supermarket checkout counter, "Paper or plastic?" should be replaced with, "Do you have your canvas shopping bag with you?"

The challenge we now face is to replace the throwaway economy with a reduce/reuse/recycle economy. The earth can no longer tolerate the pollution, the energy use, the disruption from mining, and the deforestation that the throwaway economy requires. For cities like New York, the challenge is not so much what to do with the garbage as it is how to avoid producing it in the first place.

New York recycles only 18 percent of its municipal waste. Los Angeles recycles 44 percent and Chicago 47 percent. Seattle and Minneapolis are both near 60 percent recycling rates. But even they are not close to exploiting the full potential of garbage recycling.[9]

There are many ways of shrinking the daily mountain of garbage. One is simply to ban the use of one-way beverage containers, something that Denmark and Finland have done. Denmark, for example, banned one-way soft drink containers in 1977 and beer containers in 1981. If Mayor Bloomberg wants a closer example of this approach, he need only go to Prince Edward Island in Canada, which has adopted a similar ban on one-way containers.[10]

There are other gains from reusing beverage containers. Since refillable containers are simply back-hauled to the original soft drink or brewery bottling sites by the same trucks that deliver the beverages, they reduce not only garbage but also traffic congestion, energy use, and air pollution.

We have the technologies to recycle virtually all the components of garbage. For example, Germany now gets 72 percent of its paper from recycled fiber. With glass, aluminum, and plastic, potential recycling rates are even higher.[11]

The nutrients in garbage can also be recycled by composting organic materials, including yard waste, table waste, and produce waste from supermarkets. Each year, the world mines 139 million tons of phosphate rock and 20 million tons of potash to obtain the phosphorus and potassium needed to replace the nutrients that crops remove from the soil. Urban composting that would return nutrients to the land could greatly reduce this expenditure on nutrients and the disruption caused by their mining.[12]

Yet another garbage-reducing step in this fiscally stressed situation would be to impose a tax on all throwaway products, in effect a landfill tax, so that those who use throwaway products would directly bear the cost of disposing of them. This would increase revenues while reducing garbage disposal expenditures, helping to reduce the city's fiscal deficit.

There are numerous win-win-win solutions that are

both economically attractive and environmentally desirable, and that will help avoid the unfolding public relations debacle created by the image of New York as garbage capital of the world. A response to this situation that treats the causes rather than the symptoms of garbage generation could work wonders for the city.

For additional information, see <www.earth-policy.org/Updates/Update10.htm>.

July 2002

Tax Shifting on the Rise

Bernie Fischlowitz-Roberts

Many countries have implemented taxes on environmentally destructive products and activities while simultaneously reducing taxes on social security contributions or income. The scale of tax shifting has been relatively small thus far, accounting for only 3 percent of tax revenues worldwide. It is increasingly clear, however, that countries are recognizing the power of tax systems not only for raising revenue, but also for shaping economic decisions of individuals and businesses. The German tax shift, one of the most advanced to date, illustrates how countries are modifying tax systems to reach environmental and economic objectives.[1]

Germany has implemented environmental tax reform in several stages by lowering income taxes and raising energy taxes. In 1999, taxes on gasoline, heating oils, and natural gas were increased, and a new tax on electricity was adopted. This revenue was used to decrease employer and employee contributions to the pension fund. Energy tax rises for many energy-intensive industries were substantially lower, however, reflecting concerns about international competitiveness.[2]

The second stage, which began in 2000, involved further reductions in payroll taxes and increases on those on motor fuels and electricity. Germany has shifted 2 percent of its tax burden from incomes to environmentally destructive activities. As a result, fuel sales were 5 percent lower in the first half of 2001 than in the same period in 1999. Consumption of gasoline decreased by 12 percent over the same time, and carpool agencies reported growth of 25 percent in the first half of 2000.[3]

Though Chancellor Gerhard Schroeder, concerned about the September 2002 elections, recently prevented any additional energy tax increases until 2003, his main opponent, Edward Stoiber, has pledged to continue with environmental tax reforms if elected. The agreement of Schroeder and Stoiber on the need to continue the tax shift is encouraging.[4]

The idea behind tax shifting is that raising taxes on products and activities that society wishes to discourage will encourage more environmentally friendly ways of doing business. For example, one part of the United Kingdom's environmental tax reform involved a steadily increasing fuel tax known as a fuel duty escalator, which was in effect from 1993 until 1999. As a result, fuel consumption from road transportation dropped, and the average fuel efficiency of trucks over 33 tons increased by 13 percent between 1993 and 1998. Ultra-low sulfur diesel had a lower tax rate than regular diesel, which caused its share of domestic diesel sales to jump from 5 percent in July 1998 to 43 percent in February 1999; by the end of 1999, the nation had completely converted to ultra-low sulfur diesel.[5]

The Netherlands has also implemented a series of environmental tax shifts. A general fuel tax, originally implemented in 1988 and modified in 1992, is levied on fossil fuels; the rates are based on both the carbon and the energy contents of the fuel. Between 1996 and 1998, a Regulatory Energy Tax (RET) was implemented, which taxed natural gas, electricity, fuel oil, and heating oil. Unlike the fuel tax, which was designed principally for revenue generation, the RET's goal was to change behaviors by creating incentives for energy efficiency. To maintain competitiveness, major energy users were exempted from the taxes, so this tax fell mainly on individuals.[6]

Since 60 percent of the revenue from these Dutch taxes came from households, the taxes were offset by decreasing income taxes. The 40 percent of revenue derived from businesses was recycled through three mechanisms: a reduction in employer contributions to social security, a reduction in corporate income taxes, and an increased tax exemption for self-employed people. This tax shift has caused household energy costs to increase, which has resulted in a 15-percent reduction in consumer electricity use and a 5- to 10-percent decrease in fuel usage.[7]

Finland first implemented a carbon dioxide (CO_2) tax in 1990. Between 1990 and 1998, the country's CO_2 emissions decreased by almost 7 percent. Finland's environmental taxes, like those in most countries, are far from uniform: the electricity tax is greater for households and the service sector than for industry.[8]

Sweden's experiment with tax shifting began in 1991, when it raised taxes on carbon and sulfur emissions and reduced income taxes. Manufacturing industries received exemptions and rebates from many of the environmental taxes, and as a result their tax rates were only half of those paid by households. In 2001, the government increased taxes on diesel fuel, heating oil, and electricity while lowering income taxes and social security contributions. Six percent of all government revenue in Sweden has now been shifted. This has helped Sweden reduce greenhouse gas emissions more quickly than anticipated. A political agreement between the government and the opposition required a 4-percent reduction below 1990 levels by 2012. Yet by 2000, emissions were already down 3.9 percent from 1990—in large measure due to energy taxes.[9]

A preliminary assessment of existing environmental tax shifts yields mixed results. Emissions of some taxed pollutants have decreased: some have declined absolutely, while others are lower than projected but still higher in absolute terms due to increased consumption associated with economic growth. Using price mechanisms to spur changes in consumer and producer behavior can work, but if tax rates are set too low they will not have the desired effect. The myriad exemptions given to industries, especially energy-intensive ones, in existing tax shift programs slow the restructuring. These exemptions, created out of legitimate competitiveness concerns, nonetheless slow the creation of a more effective tax system.[10]

A number of complementary policy measures can help make environmental tax shifts more effective. First, eliminating subsidies to environmentally destructive industries will help the market send the right signals. Worldwide, environmentally destructive subsidies exceed $500 billion annually. As long as government subsidies encourage activities that the taxes seek to discourage, the effectiveness of tax shifting will be limited.[11]

Second, tax harmonization within the European Union, where countries can agree on a framework of environmental tax shifts, might lessen the need for the numerous exemptions for industry that currently plague national environmental tax regimes. Even without harmonization, using border tax adjustments—where companies have environmental taxes rebated to them upon export and have domestic environmental taxes added to imports—can ensure international competitiveness.[12]

Third, when trying to guarantee equitable results of tax shifting, opting for tax refunds for lower-income citizens rather than tax exemptions preserves the incentive effect of the environmental tax. Fourth, for items whose demand does not change appreciably with small changes in price, making tax rates substantially higher—in a predictable and transparent way—will decrease consumption more than many of the limited efforts to date. Finally, expanding the tax base to encompass more products and services with deleterious environmental impacts would greatly enhance the effectiveness of tax shifting.[13]

Aviation fuel, for example, is currently tax-free worldwide, despite causing 3.5 percent of global warming. However, recent European discussions of imposing taxes on jet fuel are a promising development that might slow the projected growth in worldwide consumption by reducing air travel or by producing efficiency improvements that lower jet fuel consumption. Sweden's tax on domestic air transport, for example, prompted the one domestic airline at the time to alter the engines of its Fokker aircraft, which lowered hydrocarbon emissions by 90 percent.[14]

If properly constructed, tax shifts can help make markets work more effectively by incorporating more of the indirect costs of goods and services into their prices and by changing consumer and producer behavior accordingly. The emergence of a world-leading wind turbine industry in Denmark, for example, is one result of Danish taxes on fossil fuels and electricity, which are among the highest in the world. These measures have also spurred sales of energy-efficient appliances and encouraged other energy-saving behavior.[15]

The goal of tax restructuring is to get the market to tell the ecological truth. Thus far, tax shifts have been modest in scope and have produced positive, if modest, results. Creation of an eco-economy calls for tax shifts of much larger magnitude in order for prices to reflect their full costs and to produce the requisite changes in individual and collective behavior.

For additional information, see <www.earth-policy.org/Updates/Update14.htm>.

Notes

PART 1. THE ECONOMIC COSTS OF ECOLOGICAL DEFICITS

Deserts Invading China (pages 7–28)

1. Howard W. French, "China's Growing Deserts are Suffocating Korea," *New York Times*, 14 April 2002.
2. Ibid.
3. Ibid.
4. "Grapes of Wrath in Inner Mongolia," report from U.S. Embassy in Beijing, May 2001, at <www.usembassy-china.org.cn/english/sandt/MongoliaDust-web.htm>, viewed 6 June 2002.
5. Ibid.
6. "In Brief: Lhasa Dust Storm," *China Daily*, 29 January 2002; Wang Tao, "The Process and Its Control of Sandy Desertification in Northern China," seminar on desertification in China, Cold and Arid Regions Environmental and Engineering Research Institute, Chinese Academy of Sciences (Lanzhou, China: May 2002).
7. Guo Aibing and Jiang Zhuqing, "Airborne Dust Blankets City," *China Daily*, 21 March 2002.
8. Yang Youlin, "Dust-Sandstorms: Inevitable Consequences of Desertification—A Case Study of Desertification Disasters in the Hexi Corridor, NW China," in Yang Youlin, Victor

Squires, and Lu Qi, eds., *Global Alarm: Dust and Sandstorms from the World's Drylands* (New York: United Nations, 2001), p. 228.

9. Ibid., p. 229.

10. Ibid., p. 231.

11. Population from United Nations, *World Population Prospects: The 2000 Revision* (New York: February 2001).

12. "Grapes of Wrath in Inner Mongolia," op. cit. note 4.

13. Feng Jiaping for the State Forestry Administration, second national survey on desertification released in Beijing, cited in "Desertification Area Extends in China," *China Daily*, 29 January 2002.

14. Wang, op. cit. note 6.

15. Environmental Protection Agency cited in French, op. cit. note 1.

16. Ibid.

17. "Grapes of Wrath in Inner Mongolia," op. cit. note 4.

18. Qu cited in "China Adopts Law to Control Desertification," report from U.S. Embassy in Beijing, November 2001, at <www.usembassy-china.org.cn/sandt/desertifiction_law.htm>, viewed 6 June 2002.

19. Hong Yang and Xiubin Li, "Cultivated Land and Food Supply in China," *Land Use Policy*, vol. 17, no. 2 (2000).

20. U.N. Food and Agriculture Organization (FAO), *FAOSTAT Statistics Database*, at <www.apps.fao.org>, updated 28 May 2002.

21. "Grapes of Wrath in Inner Mongolia," op. cit. note 4.

22. Ibid.

23. Wood demand from FAO, op. cit. note 20, updated 19 December 2001.

24. Tree loss from Carmen Revenga et al., *Watersheds of the World* (Washington, DC: World Resources Institute and Worldwatch Institute, 1998); Wong Hangchang, "Deforestation and Desiccation in China: A Preliminary Study," study for

the Beijing Center for Environment and Development, Chinese Academy of Social Sciences (Beijing: 1999).

25. Economist Intelligence Unit, "China Industry: Heavy Usage, Pollution Are Hurting Water Resources, *EIU ViewsWire*, 27 February 2001; Michael Ma, "Northern Cities Sinking as Water Table Falls," *South China Morning Post*, 11 August 2001.

26. Cited in Lester R. Brown and Brian Halweil, "China's Water Shortages Could Shake World Food Security," *World Watch*, July/August 1998, pp. 11–12.

27. Water-to-grain conversion based on 1,000 tons of water for 1 ton of grain from FAO, *Yield Response to Water* (Rome: 1979), on world wheat prices from International Monetary Fund (IMF), *International Financial Statistics* (Washington, DC: various years), and on industrial water intensity in Mark W. Rosegrant, Claudia Ringler, and Roberta V. Gerpacio, "Water and Land Resources and Global Food Supply," paper presented at the 23rd International Conference of Agricultural Economists on Food Security, Diversification, and Resource Management: Refocusing the Role of Agriculture?, Sacramento, CA, 10–16 August 1997; Wang Ying, "Rice Cropped for Water," *China Daily*, 9 January 2002.

28. Ci Longjun, "Disasters of Strong Sandstorms Over Large Areas and the Spread of Land Desertification in China," in Yang, Squires, and Lu, op. cit. note 8, p. 215; Lu Qi and Ju Hongbo, "Root Causes, Processes and Consequence Analaysis of Sandstorms in Northern China in 2000," in ibid., p. 241.

29. Chen Xiwen, Deputy Director, Development Research Center of the State Council, and colleagues, discussion in Beijing with author, 16 May 2002.

30. Yang Jumping, Master Researchist, Inner Mongolia Academy of Forestry Science, Hohot, Inner Mongolia, and other local officials of the Ministry of Forestry, discussion in Hohhot with author, May 2002.

31. Data are from discussions with officials of Helin county, Inner Mongolia, 17 May 2002.

32. Ibid.

33. Wang, op. cit. note 6.

34. Ibid.

35. Author's observation confirmed by discussions with scientists at the Cold and Arid Regions Environmental and Engineering Research Institute, Lanzhou, China.

36. "China to Spend Billions on Forests," *Reuters*, 14 May 2002.

37. Shi Yuanchun, China Academy of Sciences, quoted in Frank Langfitt, "Driven by Weather, Waste, Deserts Swallowing China," *Baltimore Sun*, 20 April 2002; Jonathan Ansfield, "Sandstorms Hit China, Threaten Green Olympics Dream," *Reuters*, 21 March 2002.

38. FAO, op. cit. note 20; "Grapes of Wrath in Inner Mongolia," op. cit. note 4.

39. Langfitt, op. cit. note 37; "China Adopts Law to Control Desertification," op. cit. note 18.

40. "China Adopts Law to Control Desertification," op. cit. note 18.

41. Wang, op. cit. note 6; Asian Development Bank, *Technical Assistance to the People's Republic of China For Optimizing Initiatives to Combat Desertification in Gansu Province* (Manila, Philippines: June 2001).

42. "Grapes of Wrath in Inner Mongolia," op. cit. note 4.

43. Ibid.

44. Asian Development Bank, op. cit. note 41.

45. Grain production in 1950 from U.S. Department of Agriculture (USDA), *World Grain Database*, unpublished printout; Figure 1–1 and current levels from USDA, *Production, Supply, and Distribution*, electronic database, updated 10 May 2002.

46. USDA, *Production, Supply, and Distribution*, op. cit. note 45; IMF, op. cit. note 27.

47. Grain import dependence from USDA, *Production, Supply, and Distribution*, op. cit. note 45.

48. Data for China's trade surplus with the United States from the U.S. Department of Commerce; grain prices from *Wall Street Journal*, various issues.

Assessing the Food Prospect (pages 29–58)

1. Output in 1900 from Angus Maddison, *Monitoring the World Economy 1820–1992* (Paris: Organisation for Economic Co-operation and Development, 1995); recent growth from David Malin Roodman, "Economic Growth Falters," in Worldwatch Institute, *Vital Signs 2002* (New York: W.W. Norton & Company, 2002), pp. 58–59, and from International Monetary Fund (IMF), *World Economic Outlook* (Washington, DC: April 2002).
2. United Nations, *World Population Prospects: The 2000 Revision* (New York: February 2001).
3. Roodman, op. cit. note 1; World Bank, *World Development Report 2000/2001: Attacking Poverty* (New York: Oxford University Press, 2001), p. 3.
4. Population from United Nations, op. cit. note 2; economy from Roodman, op. cit. note 1; Population Reference Bureau, *World Population Data Sheet* (wall chart) (Washington, DC: 2001).
5. Water use from Peter H. Gleick, *The World's Water 2000–2001* (Washington, DC: Island Press, 2000), p. 52; fish from U.N. Food and Agriculture Organization (FAO), *Yearbook of Fishery Statistics: Capture Production* (Rome: various years); paper from Janet N. Abramovitz and Ashley T. Mattoon, *Paper Cuts: Recovering the Paper Landscape*, Worldwatch Paper 149 (Washington, DC: Worldwatch Institute, December 1999), p. 6; forests products and recent beef and mutton from FAO, *FAOSTAT Statistics Database*, at <apps.fao.org>, with forestry data updated 19 December 2001 and meat data updated 28 May 2002; 1950 meat demand from FAO, *1948–1985 World Crop and Livestock Statistics* (Rome: 1987).
6. Hari Eswaran, Paul Reich, and Fred Beinroth, "Global Desertification Tension Zones," in D. E. Stott, R. H. Mohtar and G. C. Steinhardt (eds.), *Sustaining the Global Farm*, selected papers from the 10th International Soil Conservation Organization Meeting held 24–29 May 1999 at Purdue University and the USDA-ARS National Soil Erosion Research Laboratory (2001), pp. 24–28.

7. Ibid.

8. Kofi A. Annan, United Nations, "Message on World Day to Combat Desertification and Drought," 17 June 2002, at <www.unccd.int/publicinfo/june17/sgmessage-eng.pdf>.

9. L. R. Oldeman, R. T. A. Hakkeling, and W. G. Sombroek, *World Map of the Status of Human-induced Soil Degradation: An Explanatory Note* (Wageningen, Netherlands: International Soil Reference and Information Centre (ISRIC), 1990).

10. Effect of topsoil loss on yields in Leon Lyles, "Possible Effects of Wind Erosion on Soil Productivity," *Journal of Soil and Water Conservation*, November/December 1975, discussed in Lester R. Brown, "Conserving Soils," in Lester R. Brown et al., *State of the World 1984* (New York: W.W. Norton & Company, 1984), pp. 62–65.

11. Rattan Lal, "Erosion-Crop Productivity Relationships for Soils of Africa," *Soil Science Society of America Journal*, May–June 1995.

12. Figure 1–2 from U.S. Department of Agriculture (USDA), *Production, Supply, and Distribution*, electronic database, updated 10 May 2002.

13. Topsoil loss from USDA, Economic Research Service (ERS), *Agri-Environmental Policy at the Crossroads: Guideposts on a Changing Landscape*, Agricultural Economic Report No. 794 (Washington, DC: January 2001), p. 16; wheat yield from USDA, op. cit. note 11.

14. USDA, op. cit. note 13; loss of topsoil from water erosion from USDA, *Summary Report: 1997 Natural Resources Inventory* (Washington, DC: December 1999, rev. December 2000), pp. 46–51; China from Chen Xiwen, Deputy Director, Development Research Center of the State Council, and colleagues, discussion with author in Beijing, 16 May 2002.

15. "Algeria to Convert Large Cereal Land to Tree-Planting," *Reuters*, 8 December 2000.

16. FAO, *The State of Food and Agriculture 1995* (Rome: 1995), pp. 174–95; wheat yields from USDA, op. cit. note 12.

17. Forest Watch Indonesia and Global Forest Watch, *The State of the Forest: Indonesia* (Bogor, Indonesia, and Washington, DC: 2002), p. 42.

18. Brazil's cerrado from "Brazil's Cerrado Land Potential," in Randall D. Schnepf, Erik N. Dohlman, and Christine Bolling, *Agriculture in Brazil and Argentina* (Washington, DC: USDA, ERS, November 2001), p. 12; Kazakhstan from FAO, op. cit. note 16; soybean production from USDA, op. cit. note 12.

19. Gleick, op cit. note 5, p. 64.

20. Population and water availability from Tom Gardner-Outlaw and Robert Engelman, *Sustaining Water, Easing Scarcity: A Second Update* (Washington, DC: Population Action International, 1997).

21. Sandra Postel, *Pillar of Sand* (New York: W.W. Norton & Company, 1999), p. 255; rule of thumb from FAO, *Yield Response to Water* (Rome: 1979); grain consumption from USDA, op. cit. note 12; population from United Nations, op. cit. note 2.

22. Water usage from Gleick, op. cit. note 5, p. 52; pumping from Postel, op. cit. note 21.

23. Table 1–4 from the following: China from Michael Ma, "Northern Cities Sinking as Water Table Falls," *South China Morning Post*, 11 August 2001; United States from Postel, op. cit. note 21, p. 77, and from Bonnie Terrell and Phillip N. Johnson, "Economic Impact of the Depletion of the Ogallala Aquifer: A Case Study of the Southern High Plains of Texas," presented at the American Agricultural Economics Association annual meeting in Nashville, TN, 8–11 August 1999; Pakistan, India, and Mexico in Tushaar Shah et al., *The Global Groundwater Situation: Overview of Opportunities and Challenges* (Colombo, Sri Lanka: International Water Management Institute, 2000); Postel, op. cit. note 21; Iran from Chenaran Agricultural Center, Ministry of Agriculture, according to Hamid Taravati, publisher, Iran, e-mail to author, 25 June 2002; Christoper Ward, "Yemen's Water Crisis," based on a lecture to the British Yemeni Society in September 2000, at <www.al-bab.com/bys/articles/ward 01.htm>, July 2001.

24. Taravati, op. cit. note 23.

25. Population from United Nations, op. cit. note 2; Yemen's water situation from Ward, op. cit. note 23; Christopher Ward, *The Political Economy of Irrigation Water Pricing in Yemen* (Sana'a, Yemen: World Bank, November 1998); Marcus Moench, "Groundwater: Potential and Constraints," in Ruth S. Meinzen-Dick and Mark W. Rosegrant, eds., *Overcoming Water Scarcity and Quality Constraints* (Washington, DC: International Food Policy Research Institute, October 2001).

26. Water table dropping 1.5 meters a year from James Kynge, "China Approves Controversial Plan to Shift Water to Drought-Hit Beijing," *Financial Times*, 7 January 2000.

27. Grain production from USDA, op. cit. note 12; irrigation from FAO, *FAOSTAT Statistics Database*, op. cit. note 5, with irrigation data updated 10 July 2001.

28. Water to grain conversion from FAO, op. cit. note 21.

29. USDA, op. cit. note 12; USDA, *World Agricultural Supply and Demand Estimate* (Washington, DC: 12 June 2002).

30. Overpumping from Postel, op. cit. note 21; population from United Nations, op. cit. note 2.

31. Water value comparison based on ratio of 1,000 tons of water for 1 ton of grain from FAO, op. cit. note 21, on global wheat prices from IMF, *International Financial Statistics* (Washington, DC: various years), and on industrial water intensity in Mark W. Rosegrant, Claudia Ringler, and Roberta V. Gerpacio, "Water and Land Resources and Global Food Supply," paper prepared for the 23rd International Conference of Agricultural Economists on Food Security, Diversification, and Resource Management: Refocusing the Role of Agriculture?, Sacramento, CA, 10–16 August 1997.

32. Grain consumption from USDA, op. cit. note 12; grain prices from IMF, op. cit. note 31; hunger and malnutrition from FAO, *The State of Food Insecurity in the World 2001* (Rome: 2001), p. 2.

33. Population from United Nations, op. cit. note 2.

34. Beef and mutton from FAO, *Crop and Livestock Statistics*, op. cit. note 5; FAO, *FAOSTAT Statistics Database*, op. cit. note 5, with meat production updated 28 May 2002; fish from FAO, *Fishery Statistics*, op. cit. note 5, and from FAO, *Aquaculture Production* (various years).

35. Figure 1–3 and data from FAO, *FAOSTAT Statistics Database*, op. cit. note 5.

36. A. Banerjee, "Dairying Systems in India," *World Animal Review*, vol. 79/2 (Rome: FAO, 1994); S. C. Dhall and Meena Dhall, "Dairy Industry—India's Strength Is in Its Livestock," *Business Line,* Internet Edition of *Financial Daily* from *The Hindu* group of publications, at <www.indiaserver.com/businessline/1997/11/07/stories/03070311.htm>, 7 November 1997; milk production data from FAO, *FAOSTAT Statistics Database*, op. cit. note 5, updated 28 May 2002.

37. China's crop residue production and use from Gao Tengyun, "Treatment and Utilization of Crop Straw and Stover in China," *Livestock Research for Rural Development*, February 2000; USDA, ERS, "China's Beef Economy: Production, Marketing, Consumption, and Foreign Trade," *International Agriculture and Trade Reports: China* (Washington, DC: July 1998), p. 28.

38. Conversion ratio for grain to beef based on Allen Baker, Feed Situation and Outlook staff, ERS, USDA, Washington, DC, discussion with author, 27 April 1992; pork conversion data from Leland Southard, Livestock and Poultry Situation and Outlook Staff, ERS, USDA, Washington, DC, discussion with author, 27 April 1992; feed-to-poultry conversion ratio derived from data in Robert V. Bishop et al., *The World Poultry Market—Government Intervention and Multilateral Policy Reform* (Washington, DC: USDA, 1990); conversion ratio for fish from USDA, ERS, "China's Aquatic Products Economy: Production, Marketing, Consumption, and Foreign Trade," *International Agriculture and Trade Reports: China* (Washington, DC: July 1998), p. 45.

39. USDA, op. cit. note 12.

40. Fish feed requirements from Rosamond L. Naylor et al., "Effect of Aquaculture on World Fish Supplies," *Nature*, 29

June 2000, p. 1019; poultry feed requirements from Bishop et al., op. cit. note 38.

41. Beef conversion from Baker, op. cit. note 38; grain to pork conversion from Southard, op. cit. note 38.

42. Aquaculture from FAO, op. cit. note 34; beef from FAO, *FAO-STAT Statistics Database*, op. cit. note 5, with meat updated 28 May 2002.

43. China's fish farms from K. J. Rana, "China," in *Review of the State of World Aquaculture*, FAO Fisheries Circular No. 886 (Rome: 1997); China's grain area from USDA, op. cit. note 11; U.S. catfish farms from USDA, ERS, National Agriculture Statistics Service, *Catfish Production* (Washington, DC: July 2000), p. 3.

44. Figure 1–4 from USDA, op. cit. note 12.

45. Soybean harvest from USDA, Foreign Agricultural Service, *Oilseeds: World Markets and Trade* (Washington, DC: May 2002).

46. Harvest area from USDA, op. cit. note 12; double cropping from Conservation Technology Information Center (CTIC), "Conservation Tillage Survey Data: Crop Residue Management 1998," CTIC Core 4 Conservation Web site, at <www.ctic.purdue.edu/Core4/CT/CT.html>, updated 19 May 2000.

47. USDA, op. cit. note 12; animal protein consumption from FAO, *FAOSTAT Statistics Database*, op. cit. note 5, updated 28 May 2002.

48. U.S. experience in USDA, op. cit. note 13; USDA, op. cit. note 14; China from Chen, op. cit. note 14.

49. USDA, Natural Resources Conservation Service, *CORE4 Conservation Practices Training Guide: The Common Sense Approach to Natural Resource Conservation* (Washington, DC: August 1999); Rolf Derpsch, "Frontiers in Conservation Tillage and Advances in Conservation Practice," in Stott, Mohtar, and Steinhardt, op. cit. note 6, pp. 248–54.

50. CTIC, "2000 United States Summary," from *2000 National Crop Residue Management Survey*, at <www.ctic.purdue.

edu/Core4/CT/ctsurvey/2000/2000USSummary.html>, updated 20 January 2002.

51. CTIC, "No-Till Adoption in the U.S.," from *2000 National Crop Residue Management Survey*, at <www.ctic.purdue.edu/Core4/CT/ctsurvey/2000/GraphNTAll.html>, updated 20 January 2002.

52. Derpsch, op. cit. note 49.

53. USDA, op. cit. note 12.

54. Sandra Postel, *Last Oasis* (New York: W.W. Norton & Company, 1997), p. 170.

55. Diversion of 70 percent from Gleick, op. cit. note 5, p. 64; Sandra Postel, "Redesigning Irrigated Agriculture," in Lester Brown et al., *State of the World 2000* (New York: W.W. Norton & Company, 2000); Sandra Postel et al., "Drip Irrigation for Small Farmers: A New Initiative to Alleviate Hunger and Poverty," *Water International*, March 2001, pp. 3–13.

56. Postel, op. cit. note 21, pp. 189–92.

57. Population projections in United Nations, op. cit. note 2.

Facing the Climate Challenge (pages 59–80)

1. National Aeronautics and Space Administration, Goddard Institute for Space Studies, "Global Temperature Anomalies in .01 C," at <www.giss.nasa.gov/data/update/gistemp/GLB.Ts.txt>, viewed 20 June 2002.

2. Ibid.

3. Intergovernmental Panel on Climate Change, *Climate Change 2001: The Scientific Basis. Contribution of Working Group I to the Third Assessment Report of the Intergovernmental Panel on Climate Change* (New York: Cambridge University Press, 2001), p. 13.

4. Timothy Egan, "Alaska, No Longer So Frigid, Starts to Crack, Burn and Sag," *New York Times*, 16 June 2002; Andes from Andrew Revkin, "A Message in Eroding Glacial Ice: Humans Are Turning Up the Heat," *New York Times*, 19 February 2001; Himalayas from Robert Marquand, "Glaciers in the

Himalayas Melting at a Rapid Rate," *Christian Science Monitor*, 5 November 1999.

5. Richard Kerr, "Will the Arctic Ocean Lose All Its Ice?" *Science*, 3 December 1999, p. 1828.

6. Dorthe Dahl-Jensen, "The Greenland Ice Sheet Reacts," *Science*, 21 July 2000, pp. 404–05.

7. Munich Re, *Topics Annual Review: Natural Catastrophes 2001* (Munich, Germany: 2002), pp. 16–17.

8. Jeremy Leggett, "The Emerging Response of the Insurance Industry to the Threat of Climate Change," *UNEP Industry and Environment*, January–March 1994, p. 41; Munich Re, op. cit. note 7.

9. Munich Re, op. cit. note 7; Doug Rekenthaler, "China Survives Fourth Yangtze Flood Crest as Fifth Begins its Journey," *Disaster Relief*, 11 August 1998; Munich Re, "Munich Re's Review of Natural Catastophes in 1998," press release (Munich: 19 December 1998); Erik Eckholm, "Chinese Leaders Vow to Mend Ecological Ways," *New York Times*, 30 August 1998.

10. Munich Re quoted in Leggett, op. cit. note 8, p. 42.

11. World Bank, *World Development Report 1999/2000* (New York: Oxford University Press, 2000), p. 100.

12. Rice exports from U.S. Department of Agriculture (USDA), Foreign Agricultural Service, *Grain: World Markets and Trade*, April 2002, p. 13; population from United Nations, *World Population Prospects: The 2000 Revision* (New York: February 2001).

13. USDA, op. cit. note 12.

14. U.S. coastal property damage from James E. Neumann et al., *Sea-level Rise & Global Climate Change: A Review of Impacts to U.S. Coasts* (Arlington, VA: Pew Center on Global Climate Change, 2000), pp. 4, 31.

15. Ibid.; coastal counties population from National Oceanic and Atmospheric Administration, *State of the Coast Report*, "Population: Distribution, Density, and Growth," at <state-of-coast.noaa.gov/bulletins/html/pop_01/national.html>, viewed 25 June 2002.

16. "Heat Wave Deaths Top 1,000," *Associated Press*, 23 May 2002; "India: Heat Toll Up to 760," *Agence France-Presse*, 22 May 2002; Islamabad from World Weather Forecast, *Washington Post*, 15 June 2002.

17. USDA, *Production, Supply, and Distribution*, electronic database, Washington, DC, updated May 2002.

18. BP, *BP Statistical Review of World Energy 2001* (London: Group Media & Publishing, June 2001).

19. Coal from ibid.

20. Table 1–8 compiled by Earth Policy Institute from BP, *BP Statistical Review of World Energy 2002* (London: Group Media & Publishing, June 2002), from American Wind Energy Association (AWEA), *Global Wind Energy Market Report* (Washington DC: March 2002), from Worldwatch Institute, *Vital Signs 2002* (New York: W.W. Norton & Company, 2002), from Paul Maycock, *PV News*, various issues; and from Geothermal Energy Association, "World Geothermal Power Up 50%, New US Boom Possible," press release (Washington, DC: 11 April 2002).

21. "Power to the Poor," *The Economist*, 10 February 2001, pp. 21–23.

22. International Geothermal Association, *Interactive World Map*, at <iga.igg.cnr.it/index.php>.

23. Hydroelectric from BP, op. cit. note 18.

24. Figure 1–5 from AWEA, op. cit. note 20, from Christopher Flavin, "Wind Energy Surges," in Worldwatch Institute, op. cit. note 20, pp. 42–43, and from *Windpower Monthly*, various issues.

25. "Winds over European Waters Harnessed for Electricity," *Environmental News Network*, 17 December 2001. According to AWEA, Kansas, North Dakota, and Texas would be able to produce 3,470 billion kilowatt-hours (kWh), exceeding the 3,087 billion kWh used by the United States in 2000, as reported by Department of Energy, Energy Information Administration (DOE, EIA); AWEA, *AWEA Wind Energy Projects Database*, at <www.awea.org/projects/index.html> and EIA Country Analysis Brief, DOE, at <www.eia.doe.gov/emeu/

cabs/usa.html>. According to Debra Lew and Jeffrey Logan, "Energizing China's Wind Power Sector," Pacific Northwest Laboratory, 2001, at <www.pnl.gov/china/ChinaWnd.htm> viewed 25 May 2001, China has at least 250 gigawatts of exploitable wind potential, roughly equal to the current installed electrical capacity in China as reported by EIA.

26. Larry Flowers, National Renewable Energy Laboratory, "Wind Power Update," at <www.eren.doe.gov/wind poweringamerica/pdfs/wpa/wpa_update.pdf>, viewed 19 June 2002; Glenn Hasek, "Powering the Future," *Industry Week*, 1 May 2000.

27. Honda and Daimler-Chrysler from Ann Job, "The Hybrids Are Coming," *Associated Press*, 12 March 2002; Ford from *Hydrogen & Fuel Cell Letter*, April 2002, at <www.hfc letter.com/letter/April02/features.html>, viewed 19 June 2002.

28. Seth Dunn, "The Hydrogen Experiment," *World Watch*, November/December 2000, pp. 14–25; Reykjavik's buses from World Business Council for Sustainable Development, "DaimlerChrysler, Shell, and Norsk Hydro: The Iceland Experiment," at <www.wbcsd.ch/casestud/iceland>, viewed 25 June 2002.

29. "BP to Build Singapore Stations for Hydrogen Cars," *Reuters*, 23 October 2001.

30. Denmark and Germany from AWEA, op. cit note 20, pp. 3–4; Navarra from Felix Avia Aranda and Ignacio Cruz Cruz, "Breezing Ahead: The Spanish Wind Energy Market," *Renewable Energy World*, May–June 2000; wind electric capacity in California from AWEA, "California Wind Energy Projects," at <www.awea.org/projects/california.html>, viewed 23 June 2002, and San Francisco's population from Census 2000, City and County of San Francisco, at <census.abag.ca.gov/counties/SanFranciscoCounty.pdf>, viewed 25 June 2002; calculation made from wind capacity of 1,671 megawatts and using conversion factor of 1 megawatt provides energy for 1,000 people, so California has enough wind capacity to produce electricity for 1.67 million people.

31. South Dakota from Jim Dehlsen, Clipper Wind, discussion with author, 30 May 2001.

32. Salem from Blair Swezey and Lori Bird, "Businesses Lead the 'Green Power' Charge," *Solar Today*, January/February 2001, p. 24.

33. Output in 1900 from Angus Maddison, *Monitoring the World Economy 1820–1992* (Paris: Organisation for Economic Co-operation and Development, 1995); recent growth from David Malin Roodman, "Economic Growth Falters," in Worldwatch Institute, op. cit. note 20, pp. 58–59, and from International Monetary Fund (IMF), *World Economic Outlook* (Washington, DC: April 2002).

34. "Annual Smoking-Attributable Mortality, Years of Potential Life Lost, and Economic Costs—United States, 1995–1999," *Morbidity and Mortality Weekly Report*, 12 April 2002.

35. Panos Institute, *Economics Forever: Building Sustainability into Economic Policy*, Panos Briefing No. 38 (London: March 2000).

36. "Flood Impact on Economy Limited," *China Daily*, 1 September 1998; Rekenthaler, op. cit. note 9; economic losses and deaths from Munich Re, op. cit. note 9 ; removal of tree cover from Carmen Revenga et al., *Watersheds of the World* (Washington, DC: World Resources Institute and Worldwatch Institute, 1998); "Forestry Cuts Down on Logging," *China Daily*, 26 May 1998; Eckholm, op. cit. note 9; Erik Eckholm, "China Admits Ecological Sins Played Role in Flood Disaster," *New York Times*, 26 August 1998; Erik Eckholm, "Stunned by Floods, China Hastens Logging Curbs," *New York Times*, 27 February 1998.

37. Damage from Munich Re, op. cit. note 9; economy from IMF, op. cit. note 33.

38. Australia in John Tierney, "A Tale of Two Fisheries," *New York Times Magazine*, 27 August 2000.

PART 2. ECO-ECONOMY INDICATORS: TWELVE TRENDS TO TRACK

Population Growing by 80 Million Annually (pages 87–90)

1. United Nations, *World Population Prospects: The 2000 Revision* (New York: February 2001).

2. Ibid.; *The Future of Fertility in Intermediate-Fertility Countries*, paper from the Expert Group Meeting on Completing the Fertility Transition (New York: U.N. Department of Economic and Social Affairs, Population Division, 11–14 March 2002).

3. United Nations, op. cit. note 1.

4. Ibid.

5. Ibid.; United Nations, *Views and Policies Concerning Population Growth and Fertility Among Governments in Intermediate Fertility Countries*, paper from the Expert Group Meeting on Completing the Fertility Transition (New York: U.N. Department of Economic and Social Affairs, Population Division, 11–14 March 2002).

6. John Bongaarts and Charles F. Westoff, *The Potential Role of Contraception in Reducing Abortion*, Working Paper 134 (New York: Population Council, 2000).

7. John Caldwell, "The Contemporary Population Challenge," paper presented at the Expert Group Meeting on Completing the Fertility Transition (New York: U.N. Department of Economic and Social Affairs, Population Division, 11–14 March 2002), p. 9; Lawrence Summers, "The Most Influential Investment," reprinted in *People and the Planet*, vol. 2, no. 1 (1993), p. 10.

8. U.N. Population Fund (UNFPA), "Meeting the Goals of the ICPD: Consequences of Resource Shortfalls up to the Year 2000," paper presented to the Executive Board of the U.N. Development Programme and the UNFPA, New York, 12–23 May 1997; UNFPA, *Population Issues Briefing Kit* (New York: Prographics, Inc., 2001), p. 23.

9. United Nations, op. cit. note 1.

10. United Nations, *World Urbanization Prospects: The 2001 Revision* (New York: March 2002).

11. Ibid.

12. United Nations, op. cit. note 1.

13. Figure 2–1 from ibid.

Economic Growth Losing Momentum (pages 91–94)

1. Figure 2–2 from David Malin Roodman, "Economic Growth Falters," in Worldwatch Institute, *Vital Signs 2002* (New York: W.W. Norton & Company, 2002), pp. 58–59.
2. International Monetary Fund (IMF), *World Economic Outlook* (Washington, DC: April 2002), p. 8.
3. Ibid.
4. Ibid., p. 31.
5. Ibid., p. 49.
6. Ibid., p. 8.
7. Ibid., p. 33.
8. Ibid.
9. Ibid.
10. Ibid., p. 42.
11. Ibid., pp. 43–44.
12. IMF, *World Economic Outlook* (Washington, DC: October 2001), pp. 46–47.
13. Unsustainable use of water from Sandra Postel, *Pillar of Sand* (New York: W.W. Norton & Company, 1999), p. 255; conversion of water to grain from U.N. Food and Agriculture Organization (FAO), *Yield Response to Water* (Rome: 1979); current grain harvest from U.S. Department of Agriculture, *World Agricultural Supply and Demand Estimate* (Washington, DC: 12 June 2002).
14. FAO, *The State of World Fisheries and Aquaculture 2000* (Rome: 2000), p. 10.

Grain Harvest Growth Slowing (pages 95–98)

1. Figure 2–3 from U.S. Department of Agriculture (USDA), *World Agricultural Supply and Demand Estimate* (Washington, DC: 12 June 2002), and from USDA, *Production, Supply, and Distribution*, electronic database, updated 10 May 2002.

2. USDA, *World Agricultural Supply and Demand Estimate*, op. cit. note 1; USDA, *Production, Supply, and Distribution*, op. cit. note 1.

3. Grain prices from International Monetary Fund, *International Financial Statistics* (Washington, DC: various years).

4. Water tables in key areas from Sandra Postel, *Pillar of Sand* (New York: W.W. Norton & Company, 1999); share of China's grain harvest from the North China Plain based on Hong Yang and Alexander Zehnder, "China's Regional Water Scarcity and Implications for Grain Supply and Trade," *Environment and Planning A*, January 2001, pp. 79–95, and on USDA, *Production, Supply, and Distribution*, op. cit. note 1; water tables falling in China and India also from International Water Management Institute, "Groundwater Depletion: The Hidden Threat to Food Security," Brief 2, at <www.cgiar.org/iwmi/intro/brief2.htm>, 2001; Bonnie L. Terrell and Phillip N. Johnson, "Economic Impact of the Depletion of the Ogallala Aquifer: A Case Study of the Southern High Plains of Texas," paper presented at the American Agricultural Economics Association annual meeting in Nashville, TN, 8–11 August 1999.

5. Ratio of 1,000 tons of water for 1 ton of grain from U.N. Food and Agriculture Organization (FAO), *Yield Response to Water* (Rome: 1979).

6. Grain imports from USDA, *Grain: World Markets and Trade* (Washington, DC: May 2002).

7. Ibid.; USDA, *Production, Supply, and Distribution*, op. cit. note 1.

8. Ibid.; 1 ton equals 1 cubic meter.

9. Water use from Peter H. Gleick, *The World's Water 2000–2001* (Washington, DC: Island Press, 2000), p. 64; grain from irrigated land in Ruth S. Meinzen-Dick and Mark W. Rosegrant, eds., "Overview," in *Overcoming Water Scarcity and Quality Constraints* (Washington, DC: International Food Policy Research Institute, October 2001).

10. USDA, *Production, Supply, and Distribution*, op. cit. note 1.

11. FAO, *FAOSTAT Statistics Database*, at <apps.fao.org>, updated 10 July 2001.

12. Fred Gale et al., "China Grain Policy at a Crossroads," *Agricultural Outlook* (Washington, DC: USDA, Economic Research Service (ERS), September 2001); Hsin-Hui Hsu and Fred Gale, coordinators, *China: Agriculture in Transition* (Washington, DC: USDA, ERS, November 2001).
13. Grain stocks from USDA, *World Agricultural Supply and Demand Estimate*, op. cit. note 1.
14. Population from United Nations, *World Population Prospects: The 2000 Revision* (New York: February 2001).

Fish Catch Leveling Off (pages 99–102)

1. Figure 2–4 from U.N. Food and Agriculture Organization (FAO), *Yearbook of Fishery Statistics: Capture Production* (Rome: various years); update from FAO, "2000 Capture Production Respect to the Previous Year, at <www.fao.org/fi/statist/snapshot/00vs99/00vs99.asp>, updated 11 March 2002; FAO, *The State of World Fisheries and Aquaculture 2000* (Rome: 2000), p. 10.
2. Reg Watson and Daniel Pauly, "Systematic Distortion in World Fisheries Catch Trends," *Nature*, 29 November 2001, pp. 534–36.
3. SeaWeb, "North Atlantic Study Reveals Food Fish Catches Have Declined by Half—Despite Tripled Fishing Effort," press release on presentation by Daniel Pauly, Andrew Rosenberg, and Reg Watson, American Association for the Advancement of Science (AAAS) annual meeting, 16 February 2002.
4. North Atlantic from ibid.; world subsidies from World Wildlife Fund (WWF), *Hard Facts, Hidden Problems: A Review of Current Data on Fishing Subsidies* (Washington, DC: October 2001), pp. ii, 3, and from FAO, *The State of Food and Agriculture 1993* (Rome: 1993), p. 58.
5. Gareth Porter, *Fisheries Subsidies and Overfishing: Towards a Structured Discussion* (Geneva: U.N. Environment Programme, February 2001), p. vii.
6. Ships from FAO, *World Fisheries and Aquaculture*, op. cit. note 1, p. 12; fuel from SeaWeb, op. cit. note 3; sustainable yields from WWF, op. cit. note 4, p. ii.

7. Aquaculture production from FAO, *Yearbook of Fishery Statistics: Aquaculture Production* (Rome: various years), and from 2000 summary tables, at <www.fao.org/fi/statist/summ tab/default.asp>, updated 2002; poultry from FAO, *FAOSTAT Statistics Database*, at <apps.fao.org>, updated 28 May 2002.

8. Rosamond L. Naylor, "Effect of Aquaculture on World Fish Supplies," *Nature*, 29 June 2000, pp. 1017–24.

9. Aquaculture production from FAO, op. cit. note 7; China from Naylor, op. cit. note 8, and from K. J. Rana, "China," in FAO, *Review of the State of World Aquaculture*, Fisheries Circular No. 886 (Rome: 1997); rice and fish polyculture from Li Kangmin, "Rice Aquaculture Systems in China: A Case of Rice-Fish Farming from Protein Crops to Cash Crops," Proceedings of the Internet Conference on Integrated Biosystems 1998, at <www.ias.unu.edu/proceedings/icibs/li/paper.htm>, viewed 5 July 2000.

10. Naylor, op. cit. note 8; Rosamond L. Naylor et al., "Nature's Subsidy to Shrimp and Salmon Farming," *Science*, 30 October 1998, pp. 883–84; Rebecca J. Goldburg, Matthew S. Elliott, and Rosamond L. Naylor, *Marine Aquaculture in the United States* (Arlington, VA: Pew Oceans Commission, 2001).

11. Benjamin Halpern, "The Impact of Marine Reserves: Do Reserves Work and Does Reserve Size Matter?" *Ecological Applications* (in press).

12. Marine Stewardship Council, at <www.msc.org>.

Forest Cover Shrinking (pages 103–07)

1. Forest worth from Robert Costanza et al., "The Value of the World's Ecosystem Services and Natural Capital," *Nature*, 15 May 1997, pp. 253–60; gross world product from International Monetary Fund, *World Economic Outlook WEO Database*, at <www.imf.org/external/pubs/ft/weo/2002/01/ index.htm>, April 2002.

2. U.N. Food and Agriculture Organization (FAO), *Forest Resources Assessment (FRA) 2000*, at <www.fao.org/forestry/ fo/fra/index.jsp>, updated 10 April 2001; historical perspective from Emily Matthews et al., *Pilot Analysis of Forest*

Ecosystems: Forest Ecosystems (Washington, DC: World Resources Institute (WRI), 2000), p. 16.

3. FAO, *State of the World's Forests 2001* (Rome: 2001), pp. 58–59.

4. Ibid., pp. 154–57; Mexico from Nick Miles, "Mexico's 'Devastating' Forest Loss, *BBC News*, 4 March 2002, and from "Mexico: Deforestation Progresses, But Not the Measures to Prevent It," *World Rainforest Movement Bulletin*, January 2002.

5. FAO, op. cit. note 3, pp. 58–59; Indonesia from Forest Watch Indonesia (FWI) and Global Forest Watch (GFW), *The State of the Forest: Indonesia* (Bogor, Indonesia, and Washington, DC: 2002), p. xi.

6. Matthews et al., op. cit. note 2, p. 3; "Reports Conclude Much of World's Remaining Intact Forests At Risk," press release (Washington, DC: WRI, 3 April 2002).

7. Matthews et al., op. cit. note 2, pp. 4–5.

8. Dave Currey et al., *Timber Trafficking: Illegal Logging in Indonesia, South East Asia and International Consumption of Illegally Sourced Timber* (London: Emerson Press, Environmental Investigation Agency and Telapak Indonesia, September 2001), p. 5; Jim Ford and Alexander Sheingauz, "Major Trends and Issues in Forests and Forestry: Globally and in Russia," presentation for Commercial Forestry in the Russian Far East: Opportunities for Sustainable Trade, Conservation and Community Development Conference, Economic Research Institute, Yuzhno-Sakhalinsk, Russia, 18–20 September 2001.

9. FAO, op. cit note 3, p. 37; Matthews et al., op. cit. note 2, p. 4.

10. FWI and GFW, op. cit. note 5, p. xii.

11. United Nations Environment Programme, *An Assessment of the Status of the World's Remaining Closed Forests* (Nairobi: 2001), p. 1.

12. Forest Stewardship Council, *Forests Certified by FSC-Accredited Bodies*, at <www.fscoax.org>, updated 30 June 2001.

Water Scarcity Spreading (pages 108–11)

1. Water demand in Peter H. Gleick, *The World's Water 2000–2001* (Washington, DC: Island Press, 2000).
2. Drying of rivers in Sandra Postel, *Pillar of Sand* (New York: W.W. Norton & Company, 1999).
3. Ibid., pp. 261–62; Jim Carrier, "The Colorado: A River Drained Dry," *National Geographic*, June 1991, pp. 4–32.
4. Lester R. Brown, "The Aral Sea: Going, Going...," *World Watch*, January/February 1991, pp. 20–27.
5. Lester R. Brown, and Brian Halweil, "China's Water Shortages Could Shake World Food Security," *World Watch*, July/August 1998, pp. 10–21.
6. Lake Chad from Michael T. Coe and Jonathan A. Foley, "Human Impacts on the Water Resources of Lake Chad," *Journal of Geophysical Research-Atmospheres*, 27 February 2001, pp. 3349–56; Hebei Province from Economist Intelligence Unit, "China Industry: Heavy Usage, Pollution Are Hurting Water Resources, *EIU ViewsWire*, 27 February 2001.
7. Water tables in key food-producing areas from Postel, op. cit. note 2; share of China's grain harvest from the North China Plain based on Hong Yang and Alexander Zehnder, "China's Regional Water Scarcity and Implications for Grain Supply and Trade," *Environment and Planning A*, vol. 33, January 2001, pp. 79–95, and on U.S. Department of Agriculture (USDA), *Production, Supply, and Distribution*, electronic database, updated 10 May 2002; water tables falling in China and India from International Water Management Institute, "Groundwater Depletion: The Hidden Threat to Food Security," Brief 2, at <www.cgiar.org/iwmi/intro/brief2.htm>, 2001; China from James Kynge, "China Approves Controversial Plan to Shift Water to Drought-Hit Beijing," *Financial Times*, 7 January 2000; Bonnie L. Terrell and Phillip N. Johnson, "Economic Impact of the Depletion of the Ogallala Aquifer: A Case Study of the Southern High Plains of Texas," paper presented at the American Agricultural Economics Association annual meeting in Nashville, TN, 8–11 August 1999.
8. Grain imports from USDA, *Grain: World Markets and Trade*

(Washington, DC: May 2002), and from USDA, op. cit. note 7.

9. Chenaran Agricultural Center, Ministry of Agriculture, according to Hamid Taravati, publisher, Iran, e-mail to author, 25 June 2002.

10. Ibid.

11. Population from United Nations, *World Population Prospects: The 2000 Revision* (New York: February 2001); Yemen's water situation from Christoper Ward, "Yemen's Water Crisis," based on a lecture to the British Yemeni Society in September 2000, at <www.al-bab.com/bys/articles/ward01.htm>, July 2001; Christopher Ward, *The Political Economy of Irrigation Water Pricing in Yemen* (Sana'a, Yemen: World Bank, November 1998); Marcus Moench, "Groundwater: Potential and Constraints," Focus 9, in Ruth S. Meinzen-Dick and Mark W. Rosegrant, eds., *Overcoming Water Scarcity and Quality Constraints* (Washington, DC: International Food Policy Research Institute, October 2001); "High and Dry: Why Yemen is Running Out of Water," *The Economist*, 30 March 2002.

12. Population and water availability from Tom Gardner-Outlaw and Robert Engelman, *Sustaining Water, Easing Scarcity: A Second Update* (Washington, DC: Population Action International, 1997).

13. Figure 2–5 from U.N. Food and Agriculture Organization, *FAOSTAT Statistics Database*, at <apps.fao.org>, updated 10 July 2001.

Carbon Emissions Climbing (pages 112–15)

1. Figure 2–6 from Seth Dunn, "Carbon Emissions Reach New High," in Worldwatch Institute, *Vital Signs 2002* (New York: W.W. Norton & Company, 2002), pp. 52–53; Intergovernmental Panel on Climate Change (IPCC), *Climate Change 2001: The Scientific Basis. Contribution of Working Group I to the Third Assessment Report of the Intergovernmental Panel on Climate Change* (New York: Cambridge University Press, 2001), p. 7.

2. Dunn, op. cit. note 1.

3. Emission source from IPCC, op. cit. note 1; projected usage from U.S. Department of Energy, Energy Information Administration, *International Energy Outlook 2002* (Washington, DC: March 2002), pp. 26, 43, 70; decline calculations from data in British Petroleum, *Statistical Review of World Energy 2001* (London: Group Media & Publishing, June 2001).
4. Carbon dioxide concentrations from Seth Dunn, *Hydrogen Futures: Toward a Sustainable Energy System*, Worldwatch Paper 157 (Washington, DC: Worldwatch Institute, August 2001), p. 25; surface temperature projections from IPCC, op. cit. note 1, p. 13.
5. Pew Center on Global Climate Change, *Climate Change: Science, Strategies, & Solutions* (Boston: Brill, 2001), p. 380.
6. Rodger Doyle, "Greenhouse Follies," *Scientific American*, April 2002, p. 29.
7. Kyoto commitments from Hermann E. Ott, "The Kyoto Protocol to the UN Framework Convention on Climate Change—Finished and Unfinished Business," at <www.wupperinst.org/Publikationen/Kyoto_Protokoll.htm>, viewed 23 May 2002; current status from U.N. Framework Convention on Climate Change, "Kyoto Protocol: Status of Ratification," at <unfccc.int/resource/kpstats.pdf>, viewed 11 June 2002.
8. American Council for an Energy-Efficient Economy, "The 2001 Carbon Scorecard: United States Seriously Lags Behind Industrialized World in Controlling Emissions," press release (Washington, DC: 13 May 2002).
9. IPCC, op. cit. note 1, p.12.
10. Fossil fuel subsidies from Dunn, op. cit. note 4, p. 10.

Global Temperature Rising (pages 116–18)

1. Figure 2–7 from National Aeronautics and Space Administration, Goddard Institute for Space Studies, "Global Temperature Anomalies in .01 C," at <www.giss.nasa.gov/data/update/gistemp/GLB.Ts.txt>, viewed 20 June 2002.
2. Ibid.

3. Ibid.

4. Intergovernmental Panel on Climate Change (IPCC), *Climate Change 2001: The Scientific Basis. Contribution of Working Group I to the Third Assessment Report of the Intergovernmental Panel on Climate Change* (New York: Cambridge University Press, 2001), p. 13.

5. Ibid.

6. World Bank, *World Development Report 1999/2000* (New York: Oxford University Press, 2000), p. 100.

7. IPCC, op. cit. note 4.

Ice Melting Everywhere (pages 119–23)

1. Intergovernmental Panel on Climate Change (IPCC), *Climate Change 2001: The Scientific Basis. Contribution of Working Group I to the Third Assessment Report of the Intergovernmental Panel on Climate Change* (New York: Cambridge University Press, 2001). Table 2–2 from the following: Lisa Mastny, "Melting of Earth's Ice Cover Reaches New High," *Worldwatch News Brief* (Washington, DC: Worldwatch Institute, 6 March 2000), updated by Earth Policy Institute with National Snow and Ice Data Center, "Antarctic Ice Shelf Collapses," at <nsidc.org/iceshelves/larsenb2002>, 19 March 2002, with "Breakaway Bergs Disrupt Antarctic Ecosystem," *Environment News Service*, 9 May 2002, and with Lonnie G. Thompson, "Disappearing Glaciers Evidence of a Rapidly Changing Earth," American Association for the Advancement of Science annual meeting proceedings, San Francisco, CA, February 2001. Additional ice melt data from University of Colorado at Boulder, "Global Sea Levels Likely to Rise Higher in 21st Century than Previous Predictions," press release (Boulder, CO: 16 February 2002), from Mark Dyurgerov, *Glacier Mass Balance and Regime: Data of Measurements and Analysis*, Occasional Paper No. 55 (Boulder, CO: Institute of Arctic and Alpine Research, University of Colorado, 2002), and from Mark F. Meier and John M. Wahr, "Sea Level is Rising: Do We Know Why?" *Proceedings of the National Academies of Science*, 14 May 2002.

2. University of Colorado at Boulder, op. cit. note 1.
3. "Alaska Examines Impacts of Global Warming," *National Geographic News*, 21 December 2001; Mastny, op. cit. note 1.
4. American Institute of Physics, "New Research Shows Mountain Glaciers Shrinking Worldwide," press release (Boston: 30 May 2001).
5. Thompson, op. cit. note 1.
6. Mastny, op. cit. note 1.
7. Thompson, op. cit. note 1.
8. National Science Foundation, Office of Polar Programs, "Ice Sheets," at <www.nsf.gov/od/opp/support/icesheet.htm>, updated March 2001.
9. National Snow and Ice Data Center, op. cit. note 1; "Melting of Antarctic Ice Shelves Accelerates," *Environment News Network*, 9 April 1999.
10. D. A. Rothrock et al., "Thinning of the Arctic Sea-Ice Cover," *Geophysical Research Letters*, 1 December 1999, pp. 3469–72; Lars H. Smedsrud and Tore Furevik, "Towards an Ice-Free Arctic?" *Cicerone*, no. 2, 2000.
11. Richard A. Kerr, "Will the Arctic Ocean Lose All Its Ice?" *Science,* 3 December 1999, p. 1828; open water from John Noble Wilford, "Ages-Old Icecap at North Pole Is Now Liquid, Scientists Find," *New York Times*, 19 August 2000.
12. W. Krabill et al., "Greenland Ice Sheet: High Elevation Balance and Peripheral Thinning," *Science*, 21 July 2000, p. 428.
13. Ibid.
14. IPCC, op. cit. note 1.

Wind Electric Generation Soaring (pages 124–28)

1. American Wind Energy Association (AWEA), *Global Wind Energy Market Report* (Washington DC: March 2002).
2. Wind from Christopher Flavin, "Wind Energy Surges," in Worldwatch Institute, *Vital Signs 2002* (New York: W.W. Norton & Company, 2002), pp. 42–43; coal from Seth Dunn, "Fos-

sil Fuel Use Inches Up," in ibid., pp. 38–39.

3. European Wind Energy Association (EWEA), "Another Record Year for European Wind Power," press release (Brussels: 20 February 2002); AWEA, op. cit. note 1; AWEA, "Wind Energy Grew Globally at Record Clip in 2001, Report Finds," press release (Washington DC: 19 March 2002).

4. "Winds over European Waters Harnessed for Electricity," *Environment News Network*, 17 December 2001. According to AWEA, Texas, North Dakota, and Kansas would be able to produce 3,470 billion kilowatt-hours (kWh), exceeding the 3,087 billion kWh used by the United States in 2000, as reported by U.S. Department of Energy, Energy Information Administration (DOE, EIA); AWEA, *AWEA Wind Energy Projects Database*, at <www.awea.org/projects/index.html>, and EIA Country Analysis Brief, DOE, at <www.eia.doe.gov/emeu/cabs/usa.html>. According to Debra Lew and Jeffrey Logan, "Energizing China's Wind Power Sector," Pacific Northwest Laboratory, 2001, at <www.pnl.gov/china/ChinaWnd.htm>, viewed 25 May 2001, China has at least 250 gigawatts of exploitable wind potential, roughly equal to the current installed electrical capacity in China as reported by EIA.

5. Figure 2–8 from Larry Flowers, National Renewable Energy Laboratory, "Wind Power Update," at <www.eren.doe.gov/windpoweringamerica/pdfs/wpa/wpa_update.pdf>, viewed 19 June 2002, and from Glenn Hasek, "Powering the Future," *Industry Week*, 1 May 2000; tax credit extension from "US Wind Power Industry Gets Tax Credit Boost," *Reuters*, 13 March 2002; new wind farms from AWEA, op. cit. note 4.

6. Ann Job, "The Hybrids Are Coming," *Associated Press*, 12 March 2002.

7. Nordex, "Nordex Asserting Itself Well in a Difficult Environment," press release (Hamburg: 27 May 2002).

8. EWEA, AWEA, and Indian Wind Turbine Manufacturers Association, "Global Windpower Conference Heralds Major Clean Energy Expansion," press release (Paris: 2 April 2002).

9. Norton Rose, "Wind Power in France," briefing paper (London: 2001), p. 4; "Argentina Will Power the Future with Wind," *Environment News Service*, 9 February 2001; "Britain

Gears Up for Offshore Wind Power," *Reuters*, 9 April 2001; China from "Wind Wire: The Month in Brief," *Windpower Monthly*, January 2002, p. 12.

10. Stateline Project and Texas from AWEA, op. cit. note 1, p. 5; South Dakota from Jim Dehlsen, Clipper Wind, discussion with author, 30 May 2001.
11. "German Onshore Wind Plant Build Seen Peaking in 02," *Reuters*, 19 June 2002.
12. AWEA, op. cit. note 1.
13. Ibid.

Bicycle Production Breaks 100 Million (pages 129–32)

1. Figure 2–9 from Bicycle Retailer and Industry News, "World Market Report" and "Fast Facts," *Industry Directory 2002* (Santa Fe, NM: Bill Communications, 2002), pp. 4–10, and from Gary Gardner, "Bicycle Production Rolls Forward," in Worldwatch Institute, *Vital Signs 2002* (New York: W.W. Norton & Company, 2002), pp. 76–77.
2. Bicycle Retailer and Industry News, op. cit. note 1.
3. Ibid.
4. Ibid.; African market from Paul Steely White, "Africa's Bike Dealers Hold Keys to Livable Cities," *Sustainable Transport*, fall 2001, pp. 24–29.
5. White, op. cit. note 4; Paul Steely White and Noah Budnick, "A New Bicycle for Africa," *Sustainable Transport*, fall 2001, pp. 26–27.
6. White and Budnick, op. cit. note 5; Afribike from <www.afribike.org>, viewed 28 May 2002.
7. White, op. cit. note 4, p. 29.
8. Road space from Todd Litman, *Evaluating Transportation Land Use Impacts* (Victoria, BC, Canada: Victoria Transport Policy Institute, 2 April 2002), pp. 9–10; electric and fuel cell bicycles from Mike Byfield, "The Fuel-Cell Bicycle is Here," *Report Newsmagazine*, 18 February 2002, p. 34.

9. Peter Newman and Jeffrey Kenworthy, *Sustainability and Cities* (Washington, DC: Island Press, 1999), pp. 206–07; City Bikes from Klaus Hildebrandt, City Bike Foundation, e-mail to author, 17 June 2002, and from <www.bycyklen.dk>.

10. Newman and Kenworthy, op. cit. note 9, p. 208; John Pucher and Christian Lefèvre, *The Urban Transport Crisis in Europe and North America* (London: Macmillan, 1996), cited in Todd Litman, *Quantifying the Benefits of Non-Motorized Transport for Achieving TDM Objectives* (Victoria, BC, Canada: Victoria Transport Policy Institute, 1 December 1999), p. 8.

11. World automobile fleet from *Ward's World Motor Vehicle Data* (Southfield, MI: Ward's Communications, 2000); China from Philip P. Pan, "Bicycles No Longer King of the Road in China," *Washington Post*, 12 March 2001.

12. Pucher and Lefèvre, op. cit. note 10.

13. Bicycle production from Gardner, op. cit. note 1; historical automobile production data compiled by Michael Renner, "Vehicle Production Declines Slightly," in Worldwatch Institute, op. cit. note 1, p. 75.

Solar Cell Sales Booming (pages 133–37)

1. Figure 2–10 derived from Paul Maycock, *PV News*, cited in Molly O. Sheehan, "Solar Cell Use Rises Quickly," in Worldwatch Institute, *Vital Signs 2002* (New York: W.W. Norton & Company, 2002), p. 45.

2. Top five producers from Michael Schmela, "Beyond Expectations: Market Survey on World Cell Production in 2001," *Photon International*, March 2002, pp. 38–39; European Photovoltaic Industry Association (EPIA) and Greenpeace, *The Solar Generation* (Washington, DC: September 2001), p. 22.

3. EPIA and Greenpeace, op. cit. note 2, pp. 18–21; Schmela, op. cit. note 2, p. 43.

4. Paul Maycock, *PV Energy Systems*, discussion with author, 24 April 2002.

5. Calculations from ibid., and from Schmela, op. cit. note 2.

6. Current costs from Sheehan, op. cit. note 1, p. 44; Robert H. Williams, "Facilitating Widespread Deployment of Wind and Photovoltaic Technologies," *Energy Foundation 2001 Annual Report*, February 2002, pp. 21–22.

7. EPIA and Greenpeace, op. cit. note 2, p. 7; Williams, op. cit. note 6, p. 22.

8. People without electricity from U.N. Development Programme, "Introduction," in *Energy After Rio: Prospects and Challenges*, at <www.undp.org/seed/energy/chapter1.html>, viewed 21 June 2002; microcredit payoffs from Robert Freling, Solar Electric Light Fund, discussion with author, 1 May 2002.

9. EPIA and Greenpeace, op. cit. note 2, p. 27, gives the figure of up to 350 kilograms of annual carbon dioxide emissions. Calculation was made of carbon's share using the atomic weights of carbon and oxygen.

10. BP pamphlet, "Solar Electricity: Brilliantly Simple," cited in Solar Century, "Solar Homes," at <www.solarcentury.co.uk/homes>, viewed 2 May 2002; calculation on emission reduction was made using the conservative assumption of saving 0.6 kilograms of carbon dioxide per kilowatt-hour of solar output, from EPIA and Greenpeace, op. cit. note 2, multiplied by U.K. electricity consumption of 333 million kilowatt-hours in 1999 (from CIA, "United Kingdom," *World Fact Book*, at <www.cia.gov/cia/publications/factbook/geos/uk.html>, viewed 28 May 2002).

11. Maria Saporta, "'Zero Energy' Homes Are Builder's Pride," *Atlanta Journal-Constitution*, 15 April 2002.

12. Sharp and industry-wide projection from "Japan Solar Cell Makers to Boost Production," *Reuters*, 6 May 2002.

Part 3. Eco-Economy Updates

U.S. Farmers Double Cropping Corn and Wind Energy
(pages 143–47)

1. For corn, calculations by author from John Dittrich, American Corn Growers Association, "Major Crops: A 27-Year History with Inflation Adjustments," *Key Indicators of the U.S.*

Farm Sector (Washington, DC: January 2002); wind royalties from Union of Concerned Scientists, "Farming the Wind: Wind Power and Agriculture," at <www.ucsusa.org/energy/fact_wind.html>.

2. Cost reduction history from Glenn Hasek, "Powering the Future," *Industry Week*, 1 May 2000.

3. According to American Wind Energy Association (AWEA), Kansas, North Dakota, and Texas would be able to produce 3,470 billion kilowatt-hours (kWh), exceeding the 3,087 billion kWh used by the United States in 2000, as reported by U.S. Department of Energy (DOE), Energy Information Administration (EIA); AWEA, *AWEA Wind Energy Projects Database*, at <www.awea.org/projects/index.html> and EIA Country Analysis Brief, DOE, at <www.eia.doe.gov/emeu/cabs/usa.html>.

4. Beef from author's personal experience with ranches in southern Wyoming and northern Colorado; wheat from Dittrich, op. cit. note 1.

5. Calculation from Tom Gray, AWEA, e-mail to author, 12 June 2002.

6. "BTM Predicts Continued Growth for Wind Industry," *Renewable Energy Report (Financial Times)*, May 2001, p. 8; figure of 60 percent is based on listed market shares of top wind turbine suppliers; Tom Gray, "Wind is Getting Stronger and is On Course for the Next Decade," *Renewable Energy World*, May 1999.

7. Ford cited in David Bjerklie et al., "Look Who's Trying to Turn Green," *Time*, 9 November 1998.

8. Public Service Company of Colorado, "Electricity Generated by the Wind in Colorado," at <www.psco.com/solutions/wind source.asp>.

9. Minnesota from Minnesota Wind Energy Factsheet, at <www.me3.org/projects/seed/windfact.html>; Texas from DOE, Energy Efficiency and Renewable Energy Network, State Energy Alternatives, "Policy Case Study for Texas," at <www.eren.doe.gov/state_energy/policy_casestudies_texas.cfm>.

10. DOE, "Energy Secretary Richardson Directs Department to

Buy More 'Green Power,'" press release (Washington, DC: 20 April 2000).

11. AWEA, *Global Wind Energy Market Report 2000*, at <www.awea.org/faq/global2000.html>, viewed 25 June 2001; Denmark from Christopher Flavin, Worldwatch Institute, *Vital Signs 2001*, press briefing, Washington, DC, 24 May 2001; Germany from AWEA, *Wind Energy Press Background Information* (Washington, DC: February 2001), and from Christian Hinsch, "Wind Power Flying Even Higher," *New Energy*, February 2001, pp. 14–20; Navarra from Felix Avia Aranda and Ignacio Cruz Cruz, "Breezing Ahead: The Spanish Wind Energy Market," *Renewable Energy World*, May–June 2000; Debra Law and Jeffrey Logan, "Energizing China's Wind Power Sector," Pacific Northwest Laboratory, 2001, at <www.pnl.gov/china/ChinaWnd.htm>, viewed 25 May 2001.
12. Author's observations in traveling through countries.

The Rise and Fall of the Global Climate Coalition
(pages 148–52)

1. William Drozdiak, "U.S. Firms Become 'Green' Advocates," *Washington Post*, 24 November 2000.
2. PR Watch, "Impropaganda Review: Global Climate Coalition," <www.prwatch.org/improp/gcc.html>.
3. John Browne, Chief Executive, BP, speech delivered at Stanford University, Stanford, CA, 19 May 1997.
4. Ibid.; Martha M. Hamilton, "Shell Leaves Coalition That Opposes Global Warming Treaty," *Washington Post*, 22 April 1998.
5. Ford cited in David Bjerklie et al., "Look Who's Trying to Turn Green," *Time*, 9 November 1998.
6. "Ford Motor Co. Leaves Anti-Kyoto Coalition," *Environment News Service*, 7 December 1999.
7. Keith Bradsher, "Ford Announces Its Withdrawal From Global Climate Coalition," *New York Times*, 7 December 1999; David Goodman, "GM Joins DaimlerChrysler, Ford, Quits Global Warming Lobby Group," *Associated Press*, 14 March 2000; "Texaco Leaving Anti-Global Warming Treaty Group,"

Reuters, 29 February 2000; statement of Daniel Becker in Sierra Club, "Sierra Club Applauds General Motors' Exit from Global Warming Front Group," press release (Washington, DC: 14 March 2000).

8. Pew Center on Global Climate Change, Business Environmental Leadership Council, "Joint Statement of the Business Environmental Leadership Council," at <www.pewclimate.org>.

9. Pew Center on Global Climate Change, Business Environmental Leadership Council, "BP Amoco," at <www.pewclimate.org/belc/amoco.cfm>.

10. Dupont will cut emissions by 65 percent by 2010, according to their position statement, "Global Climate Change" (Wilmington, DE: 5 June 2001).

11. Seth Dunn, "The Hydrogen Experiment," *World Watch*, November/December 2000, p. 21.

12. Seth Dunn, "Fossil Fuel Use in Flux," in Lester R. Brown et al., *Vital Signs 2000* (New York: W.W. Norton & Company, 2000), pp. 52–53.

13. Lisa Mastny, "Melting of Earth's Ice Cover Reaches New High," *Worldwatch News Brief* (Washington, DC: 6 March 2000); John Noble Wilford, "Ages-Old Icecap at North Pole Is Now Liquid, Scientists Find," *New York Times*, 19 August 2000; 50-year projection in Lars H. Smedsrud and Tore Furevik, "Towards an Ice-Free Arctic?" *Cicerone*, no. 2, 2000.

14. U.S. Department of Agriculture, Economic Research Service, "Cigarette Price Increase Follows Tobacco Pact," *Agricultural Outlook*, January–February 1999.

15. PR Watch, op. cit. note 2.

16. Michael Bowlin, speech to Cambridge Energy Research Associates, 18th annual meeting, 9 February 1999.

Climate Change Has World Skating on Thin Ice (pages 153–57)

1. John Noble Wilford, "Ages-Old Icecap at North Pole Is Now Liquid, Scientists Find," *New York Times*, 19 August 2000.

2. Fifty-year projection in Lars H. Smedsrud and Tore Furevik, "Towards an Ice-Free Arctic?" *Cicerone*, no. 2, 2000; W. Krabill et al., "Greenland Ice Sheet: High Elevation Balance and Peripheral Thinning," *Science*, 21 July 2000, p. 428.
3. Lisa Mastny, "Melting of Earth's Ice Cover Reaches New High," *Worldwatch News Brief* (Washington, DC: 6 March 2000); Wilford, op. cit. note 1.
4. Krabill et al., op. cit. note 2; usable flow of the Nile River from Sandra Postel, *Pillar of Sand* (New York: W.W. Norton & Company, 1999), pp. 71, 146.
5. "Melting of Antarctic Ice Shelves Accelerates," *Environment News Network*, 9 April 1999.
6. Ibid.
7. Mastny, op. cit. note 3.
8. Ibid.
9. Ibid.
10. Fossil-fuel-related carbon emissions figure from Seth Dunn, "Carbon Emissions Continue Decline," in Worldwatch Institute, *Vital Signs 2001* (New York: W.W. Norton & Company, 2001), p. 53.
11. National Aeronautics and Space Administration, Goddard Institute for Space Studies, "Global Temperature Anomalies in .01 C," at <www.giss.nasa.gov/data/update/gistemp>, viewed 8 June 2001; Seth Dunn, "Global Temperature Steady," in Worldwatch Institute, op. cit. note 10, pp. 50–51.
12. Christopher B. Field et al., *Confronting Climate Change in California: Ecological Impacts on the Golden State* (Cambridge, MA: Union of Concerned Scientists, 1999), pp. 2–3, 10.
13. Mastny, op. cit. note 3; Dorthe Dahl-Jensen, "The Greenland Ice Sheet Reacts," *Science*, 21 July 2000, p. 404–05.
14. World Bank, *World Development Report 1999/2000* (New York: Oxford University Press, 2000), p. 100.
15. Seth Dunn, "The Hydrogen Experiment," *World Watch*, November/December 2000, pp. 14–25.

OPEC Has World Over a Barrel Again (pages 158–63)

1. Neela Banerjee, "As Prices Rise, Nations Ask for More Oil," *New York Times*, 8 September 2000.
2. Kenneth Bredemeier, "Oil Prices Hit A 10-Year High," *Washington Post*, 7 September 2000.
3. Trade in oil from BP Amoco, *BP Amoco Statistical Review of World Energy 2000* (London: Group Media & Publishing, June 2000); grain from U.S. Department of Agriculture, *Production, Supply, and Distribution*, electronic database, Washington, DC, updated August 2000.
4. Wheat and oil prices from International Monetary Fund, *International Financial Statistics* (Washington, DC: various years).
5. Ibid.
6. Ibid.; oil imports from BP Amoco, op. cit. note 3.
7. John Noble Wilford, "Ages-Old Icecap at North Pole Is Now Liquid, Scientists Find," *New York Times*, 19 August 2000; Lisa Mastny, "Melting of Earth's Ice Cover Reaches New High," *Worldwatch News Brief* (Washington, DC: 6 March 2000); 50-year projection in Lars H. Smedsrud and Tore Furevik, "Towards an Ice-Free Arctic?" *Cicerone*, no. 2, 2000.
8. Mastny, op. cit. note 7; Dorthe Dahl-Jensen, "The Greenland Ice Sheet Reacts," *Science*, 21 July 2000, p. 404–05.
9. Lester R. Brown, "The Acceleration of Change," in Lester R. Brown et al., *Vital Signs 2000* (New York: W.W. Norton & Company, 2000), p. 18.
10. International Energy Agency, "Japan: Overview of Renewable Energy Policy," in *Renewable Energy Policy in IEA Counties, Volume 2: Country Reviews* (Paris: 1998).
11. Denmark from Christopher Flavin, "Wind Power Booms," in Brown et al., op. cit. note 9, pp. 56–57; Germany from American Wind Energy Association (AWEA), *Wind Energy Press Background Information* (Washington, DC: February 2001), and from Christian Hinsch, "Wind Power Flying Even Higher," *New Energy*, February 2001, pp. 14–20; Navarra from Felix Avia Aranda and Ignacio Cruz Cruz, "Breezing Ahead:

The Spanish Wind Energy Market," *Renewable Energy World*, May–June 2000.

12. According to AWEA, Kansas, North Dakota, and Texas would be able to produce 3,470 billion kilowatt-hours (kWh), exceeding the 3,087 billion kWh used by the United States in 2000, as reported by U.S. Department of Energy (DOE), Energy Information Administration (EIA).

13. Cost of Saudi Arabian oil production from DOE/EIA, "Saudi Arabia," at <www.eia.doe.gov/emeu/cabs/saudi.html>.

14. John J. Fialka, "Clinton Seeks Saudi Help on Oil Output," *Wall Street Journal*, 8 September 2000.

Wind Power: The Missing Link in the Bush Energy Plan
(pages 164–68)

1. National Energy Policy Development Group, *National Energy Policy* (Washington, DC: U.S. Government Printing Office, May 2001).

2. Coal consumption from John Pomfret, "Research Casts Doubt on China's Pollution Claim," *Washington Post*, 15 August 2001.

3. Christopher Flavin, "Wind Energy Growth Continues," in Worldwatch Institute, *Vital Signs 2001* (New York: W.W. Norton & Company, 2001), pp. 44–45; American Wind Energy Association (AWEA), "President's Energy Plan is Useful First Step, Wind Energy Association Says," press release (Washington, DC: 17 May 2001).

4. AWEA, "US Installed Capacity (MW) 1981–2001," at <www.awea.org/faq/instcap.html>, viewed 25 June 2001; "World's Largest Wind Plant to Energize the West," PacifiCorp and FPL Energy, press release (Salt Lake City, UT, and Juno Beach, FL: 10 January 2001).

5. George Darr, "Astonishing Number of Wind Proposals Blows into BPA," Bonneville Power Administration, press release (Portland, OR: 26 April 2001).

6. Jim Dehlsen, Clipper Wind, discussion with author, 30 May 2001.

7. Cost reduction history from Glenn Hasek, "Powering the Future," *Industry Week,* 1 May 2000.

8. According to AWEA, Kansas, North Dakota, and Texas would be able to produce 3,470 billion kilowatt-hours (kWh), exceeding the 3,087 billion kWh used by the United States in 2000, as reported by U.S. Department of Energy (DOE), Energy Information Administration (EIA); AWEA, *AWEA Wind Energy Projects Database*, at <www.awea.org/projects/index.html> and EIA Country Analysis Brief, DOE, at <www.eia.doe.gov/emeu/cabs/usa.html>. According to Debra Lew and Jeffrey Logan, "Energizing China's Wind Power Sector," Pacific Northwest Laboratory, 2001, at <www.pnl.gov/china/ChinaWnd.htm>, China has at least 250 gigawatts of exploitable wind potential, roughly equal to the current installed electrical capacity in China as reported by EIA.

9. Denmark from Christopher Flavin, Worldwatch Institute, *Vital Signs 2001* press briefing, Washington, DC, 24 May 2001; Germany from AWEA, *Wind Energy Press Background Information* (Washington, DC: February 2001), and from Christian Hinsch, "Wind Power Flying Even Higher," *New Energy*, February 2001, pp. 14–20; Navarra from Felix Avia Aranda and Ignacio Cruz Cruz, "Breezing Ahead: The Spanish Wind Energy Market," *Renewable Energy World*, May–June 2000.

10. Dominique Magada, "France Sets Ambitious Target for Renewable Power," *Reuters*, 10 December 2000; Argentina from "Under Spanish Proposal, 15 Percent of Total Would be Eolic Energy," *Agencia EFE*, 7 February 2001; "UK Makes Leap into Offshore Wind Big Time," *Renewable Energy Report (Financial Times)*, May 2001.

11. European Wind Energy Association, "Wind Energy in Europe," at <www.ewea.org/src/europe.htm>.

12. Lester R. Brown, "U.S. Farmers Double-Cropping Corn and Wind Energy," *Earth Policy Alert* (Washington, DC: Earth Policy Institute, 7 June 2001).

13. "DaimlerChrysler Unveils Fuel Cell Vehicle," *Environmental News Network*, 18 March 1999; "Honda Has New Fuel-Cell Car, Toyota Expands Hybrids," *Reuters*, 29 September 2000;

Ford cited in David Bjerklie et al., "Look Who's Trying to Turn Green," *Time*, 9 November 1998.

14. Intergovernmental Panel on Climate Change, *Climate Change 2001: The Scientific Basis. Contribution of Working Group I to the Third Assessment Report of the Intergovernmental Panel on Climate Change* (New York: Cambridge University Press, 2001).

Population Growth Sentencing Millions to Hydrological Poverty (pages 169–73)

1. World population growth from United Nations, *World Population Prospects: The 2000 Revision* (New York: February 2001).
2. Ibid.
3. Sandra Postel, *Pillar of Sand* (New York: W.W. Norton & Company, 1999); rule of thumb from U.N. Food and Agriculture Organization (FAO), *Yield Response to Water* (Rome: 1979); U.S. Department of Agriculture (USDA), *Production, Supply, and Distribution*, electronic database, Washington, DC, updated May 2001.
4. USDA, op. cit. note 3.
5. Irrigation water information from World Resources Institute (WRI), *World Resources 2000–2001* (Washington, DC: 2001), p. 64; calculation based on 1,000 tons of water for 1 ton of grain from FAO, op. cit. note 3, on global wheat prices from International Monetary Fund, *International Financial Statistics* (Washington, DC: various years), and on industrial water intensity in Mark W. Rosegrant, Claudia Ringler, and Roberta V. Gerpacio, "Water and Land Resources and Global Food Supply," paper prepared for the 23rd International Conference of Agricultural Economists on Food Security, Diversification, and Resource Management: Refocusing the Role of Agriculture?, Sacramento, CA, 10–16 August 1997.
6. Average grain consumption per person derived from USDA, op. cit. note 3, and from United Nations, op. cit. note 1.
7. WRI, op. cit. note 5, p. 274.

8. FAO, op. cit. note 3.

9. USDA, op. cit. note 3.

10. Ibid.; this shows grain imports alone into the region of over 63 million tons, equivalent to 63 billion tons (63 billion cubic meters) of water, nearly the usable flow of the Nile River reported in Postel, op. cit. note 3, p. 146.

11. Total harvest from USDA, op. cit. note 3.

Africa Is Dying—It Needs Help (pages 174–79)

1. Joint United Nations Programme on HIV/AIDS (UNAIDS), *Report on the Global HIV/AIDS Epidemic* (Geneva: June 2000).

2. Ibid.

3. Ibid.

4. Ibid.

5. "AIDS, Diseases to Leave 44 Million Orphans by 2010," *Reuters*, 13 July 2000.

6. UNAIDS, op. cit. note 1, p. 29; university study from Prega Govender, "Shock AIDS Test Result at Varsity," (Johannesburg) *Sunday Times*, 25 April 1999; "South Africa: University Finds 25 Percent of Students Infected," *Kaiser Daily HIV/AIDS Report*, 27 April 1999.

7. UNAIDS, op. cit. note 1, pp. 32–33.

8. Ibid.

9. Ibid.

10. Ibid.

11. Ibid.

HIV Epidemic Restructuring Africa's Population (pages 180–84)

1. Joint United Nations Programme on HIV/AIDS (UNAIDS), *Report on the Global HIV/AIDS Epidemic* (Geneva: June 2000).

2. Ibid.

3. Ibid.

4. Ibid.

5. Ibid.

6. Desmond Cohen, *Socio-Economic Causes and Consequences of the HIV Epidemic in Southern Africa: A Case Study of Namibia*, Issues Paper No. 31 (New York: United Nations Development Programme, HIV and Development Programme, 1998).

7. Elizabeth Pisani, *Data and Decision-making: Demography's Contribution to Understanding AIDS in Africa*, Policy and Research Paper No. 14 (Paris: International Union for the Scientific Study of Population, 1998).

8. UNAIDS, op. cit. note 1, p. 11.

9. Ibid., p. 48.

10. George Tseo, "The Greening of China," *Earthwatch*, May/June 1992.

11. "AIDS, Diseases to Leave 44 Million Orphans by 2010," *Reuters*, 13 July 2000.

12. Cohen, op. cit. note 6; Richard Ingham, "Demographic Effect of AIDS South of Sahara Will Be Like Black Death," *Agence France-Presse*, 10 July 2000.

13. UNAIDS, op. cit. note 1, pp. 32–33.

Obesity Threatens Health in Exercise-Deprived Societies
(pages 185–89)

1. William H. Dietz, "Battling Obesity: Notes from the Front," National Center for Chronic Disease Prevention and Health Promotion, *Chronic Disease Notes & Reports*, winter 2000, p. 2; Ali H. Mokdad et al., "The Continuing Epidemic of Obesity in the United States," *Journal of the American Medical Association*, 4 October 2000, p. 1650.

2. National Center for Health Statistics, Centers for Disease Control and Prevention (CDC), "Prevalence of Overweight

and Obesity Among Adults," at <www.cdc.gov/nchs/products/pubs/pubd/hestats/obese/obse99.htm>, 11 December 2000; Gary Gardner and Brian Halweil, *Underfed and Overfed: The Global Epidemic of Malnutrition*, Worldwatch Paper 150 (Washington, DC: Worldwatch Institute, March 2000), p. 11; Peter G. Kopelman, "Obesity as a Medical Problem," *Nature*, 6 April 2000, p. 636; Barry M. Popkin and Colleen M. Doak, "The Obesity Epidemic is a Worldwide Phenomenon," *Nutrition Reviews*, April 1998, pp. 106–14.

3. Kopelman, op. cit. note 2, p. 636; World Health Organization (WHO), *Obesity: Preventing and Managing the Global Epidemic, Report of a WHO Consultation on Obesity* (Geneva: 1997).

4. WHO, op. cit. note 3.

5. National Center for Chronic Disease Prevention and Health Promotion, "Preventing Obesity Among Children," *Chronic Disease Notes & Reports*, winter 2000, p. 1.

6. Barry M. Popkin, "Urbanization and the Nutrition Transition," *Achieving Urban Food and Nutrition Security in the Developing World, A 2020 Vision for Food, Agriculture, and the Environment*, Focus 3, Brief 7 (Washington, DC: International Food Policy Research Institute, August 2000).

7. Gardner and Halweil, op. cit. note 2, p. 11; Kopelman, op. cit. note 2, p. 635.

8. Kopelman, op. cit. note 2, p. 635–43; Ron Winslow, "Why Fitness Matters," *Wall Street Journal*, 1 May 2000.

9. Kopelman, op. cit. note 2, p. 635.

10. Deaths from smoking from CDC, *Targeting Tobacco Use: The Nations' Leading Cause of Death* (Washington, DC: 2000); cigarette consumption from U.S. Department of Agriculture (USDA), Foreign Agricultural Service, *World Cigarette Electronic Database*, December 1999, and from USDA, Economic Research Service, *Tobacco: Situation and Outlook Report* (Washington, DC: April 2001).

11. Winslow, op. cit. note 8; Judy Putnam and Shirley Gerrior, "Trends in the U.S. Food Supply, 1970–97," in Elizabet Frazao, ed., *America's Eating Habits: Changes and Consequences*

(Washington, DC: USDA, Economic Research Service, May 1999), p. 152.

12. Winslow, op. cit. note 8.

13. Kopelman, op. cit. note 2, p. 638.

14. Ibid.

15. J.M. Friedman, "Obesity in the New Millennium," *Nature*, 6 April 2000, pp. 632–34.

16. Denise Grady, "Doctor's Review of Five Deaths Raises Concern About the Safety of Liposuction," *New York Times*, 13 May 1999.

Iran's Birth Rate Plummeting at Record Pace (pages 190–94)

1. Figure 3–1 from Central Budget and Planning Organization, and Statistics and Registration Administration of Iran, cited by Farzaneh Bahar, Former General Director of Family Planning in Iran's Khorasan state, 23 December 2001, e-mail to author; Abubakar Dungus, "Iran's Other Revolution," *Populi*, September 2000.

2. Homa Hoodfar and Samad Assadpour, "The Politics of Population Policy in the Islamic Republic of Iran," *Studies in Family Planning*, March 2000, pp. 19–34.

3. Khomeini quoted in Doug Schwartz, "Iran: Islam Embraces Contraception," *ForeignWire.com*, 18 July 1998.

4. United Nations, *World Population Prospects: The 2000 Revision* (New York: February 2001).

5. Farzaneh Roudi, "Iran's Revolutionary Approach to Family Planning," *Population Today*, July/August 1999, p. 4.

6. "Law of 23 May 1993 Pertaining to Population and Family Planning," *National Report on Population, the Islamic Republic of Iran* (Tehran, Iran: Government of Iran, 1994), pp. 20–21, at <cyber.law.harvard.edu/population/policies/IRAN.htm>.

7. Akbar Aghajanian and Amir H. Mehryar, "Fertility Transition in the Islamic Republic of Iran: 1976–1996," *Asia-Pacific Pop-*

ulation Journal, vol. 14, no. 1 (1999), pp. 21–42; Population Reference Bureau (PRB), *2001 World Population Data Sheet*, wall chart (Washington, DC: 2001).

8. Eighty percent from Roudi, op. cit. note 5; Dungus, op. cit. note 1.

9. Robin Wright, "Iran's New Revolution," *Foreign Affairs*, January/February 2000.

10. Roudi, op. cit. note 5; Schwartz, op. cit. note 3.

11. Literacy from PRB, "Iran: Demographic Highlights," fact sheet (Washington, DC: 2001); school enrollment from World Bank, *World Development Indicators 2000* (Washington, DC: 2000); television from Roudi, op. cit. note 5, p. 5.

12. Absolute water scarcity from David Seckler, David Molden, and Randolph Barker, "Water Scarcity in the Twenty-First Century," *International Water Management Institute Water Brief 1* (Sri Lanka, March 1999); "Thirst Grips Half the Population of Iran," *Environment News Service*, 4 August 2000.

13. U.S. Department of Agriculture, *Production, Supply, and Distribution*, electronic database, updated November 2001.

14. Projection for 2008 from United Nations, op. cit. note 4.

15. PRB, op. cit. note 7.

Paving the Planet: Cars and Crops Competing for Land
(pages 195–99)

1. Figure 3–2 from Michael Renner, "Vehicle Production Increases," in Lester R. Brown et al., *Vital Signs 2000* (New York: W.W. Norton & Company, 2000), pp. 86–87.

2. Calculations for paved area by Janet Larsen, Earth Policy Institute, using U.S. Department of Transportation, Federal Highway Administration (FHWA), *Highway Statistics 1999* (Washington, DC: 2001); Mark Delucchi, "Motor Vehicle Infrastructure and Services Provided by the Public Sector," cited in Todd Litman, *Transportation Land Valuation* (Victoria, B.C., Canada: Victoria Transport Policy Institute, November 2000), p. 4; *Ward's World Motor Vehicle Data* (Southfield,

MI: Ward's Communications, 2000); Jeffrey Kenworthy, Associate Professor in Sustainable Settlements, Institute for Sustainability and Technology Policy, Murdoch University, Australia, e-mail message to author; David Walterscheid, FHWA Real Estate Office, discussion with author.

3. Larsen, op. cit. note 2.

4. Ibid.; grain area from U.S. Department of Agriculture (USDA), *Production, Supply, and Distribution*, electronic database, updated January 2001.

5. Automobile production from *Ward's World Motor Vehicle Data*, op. cit. note 2; population from United Nations, *World Population Prospects: The 2000 Revision* (New York: February 2001).

6. Larsen, op. cit. note 2; population from United Nations, op. cit. note 5.

7. Larsen, op. cit. note 2; economy from International Monetary Fund, *World Economic Outlook* (Washington, DC: October 1999).

8. Larsen, op. cit. note 2; grain area from USDA, op. cit. note 4.

9. Population from United Nations, op. cit. note 5; vehicle fleet from *Ward's World Motor Vehicle Data*, op. cit. note 2.

10. Ding Guangwei and Li Shishun, "Analysis of Impetuses to Change of Agricultural Land Uses in China," *Bulletin of the Chinese Academy of Sciences*, vol. 13, no. 1 (1999).

11. Ibid.

12. Population from United Nations, op. cit. note 5.

Dust Bowl Threatening China's Future (pages 200–04)

1. "China Dust Storm Strikes USA," *NOAA News* (National Oceanic and Atmospheric Administration), 18 April 2001; Ann Schrader, "Latest Import From China: Haze," *Denver Post*, 18 April 2001.

2. "Drought Promotes Sandstorms in North China," *People's Daily*, 10 March 2001.

3. Dust storms in China from National Center for Atmospheric Research, "Scientists, Ships, Aircraft to Profile Asian Pollution and Dust." press release (Boulder, CO: 20 March 2001); U.S. Dust Bowl from William K. Stevens, "Great Plains or Great Desert? The Sea of Dunes Lies in Wait," *New York Times*, 28 May 1996.

4. Chang Jae-soon, "Korea, Japan to Cope with 'Yellow Dust'," *Korea Herald*, 7 September 2000; BBC Monitoring, "South Korea, China, Japan to Launch Joint Environmental Campaign," *Yonhap News Agency*, 27 March 2001.

5. Population from United Nations, *World Population Prospects: The 2000 Revision* (New York: February 2001).

6. Hong Yang and Xiubin Li, "Cultivated Land and Food Supply in China," *Land Use Policy*, vol. 17, no. 2 (2000).

7. Robert Henson, Steve Horstmeyer, and Eric Pinder, "The 20th Century's Top Ten U.S. Weather and Climate Events," *Weatherwise*, November/December 1999, pp. 14–19.

8. Livestock data from U.N. Food and Agriculture Organization (FAO), *FAOSTAT Statistics Database*, at <apps.fao.org>, updated 2 May 2001.

9. Erik Eckholm, "Chinese Farmers See a New Desert Erode Their Way of Life," *New York Times*, 30 July 2000.

10. Ibid.

11. Economist Intelligence Unit, "China Industry: Heavy Usage, Pollution Are Hurting Water Resources," *EIU ViewsWire*, 27 February 2001.

12. Wang Hongchang, "Deforestation and Desiccation in China: A Preliminary Study," study for the Beijing Center for Environment and Development, Chinese Academy of Social Sciences, 1999.

13. Micael C. Runnström, "Is Northern China Winning the Battle Against Desertification?" *Ambio*, December 2000, pp. 468–76.

14. Wood used for fuel from FAO, op. cit. note 8.

15. Calculation from Tom Gray, American Wind Energy Association, e-mail to author, 12 June 2002.

Worsening Water Shortages Threaten China's Food Security (pages 205–09)

1. Michael Ma, "Northern Cities Sinking as Water Table Falls," *South China Morning Post*, 11 August 2001; share of China's grain harvest from the North China Plain based on Hong Yang and Alexander Zehnder, "China's Regional Water Scarcity and Implications for Grain Supply and Trade," *Environment and Planning A*, vol. 33 (2001), and on U.S. Department of Agriculture (USDA), *Production, Supply and Distribution*, electronic database, updated September 2001.
2. Ma, op. cit. note 1.
3. Ibid.
4. Ibid.
5. World Bank, *China: Agenda for Water Sector Strategy for North China* (Washington, DC: April 2001), pp. vii, xi.
6. Hong and Zehnder, op. cit. note 1, p. 85.
7. Lester R. Brown and Brian Halweil, "China's Water Shortages Could Shake World Food Security," *World Watch*, July/August 1998, pp. 11–12.
8. Economist Intelligence Unit, "China Industry: Heavy Usage, Pollution Are Hurting Water Resources," *EIU ViewsWire*, 27 February 2001.
9. World Bank, op. cit. note 5; Zhang Qishun and Zhang Xiao, "Water Issues and Sustainable Social Development in China," *Water International*, vol. 20 (1995), pp. 122–28.
10. Population projection from United Nations, *World Population Prospects: The 2000 Revision* (New York: February 2001); water demand from Albert Nyberg and Scott Rozelle, *Accelerating China's Rural Transformation* (Washington, DC: World Bank, 1999).
11. Calculation based on 1,000 tons of water for 1 ton of grain from U.N. Food and Agriculture Organization (FAO), *Yield Response to Water* (Rome: 1979), on world wheat prices from International Monetary Fund, *International Financial Statistics* (Washington, DC: various years), and on industrial water

intensity in Mark W. Rosegrant, Claudia Ringler, and Roberta V. Gerpacio, "Water and Land Resources and Global Food Supply," paper presented at the 23rd International Conference of Agricultural Economists on Food Security, Diversification, and Resource Management: Refocusing the Role of Agriculture?, Sacramento, CA, 10–16 August 1997.

12. For more information see Fred Gale, *China's Food and Agriculture: Issues for the 21st Century* (Washington, DC: Economic Research Service, April 2002).
13. Ibid.
14. Grain harvest from USDA, op. cit. note 1.
15. Ibid.
16. Ibid.
17. John Wade and Zhang Jianping, *China: Grain and Feed Grain Update* (Beijing, USDA Foreign Agricultural Service, 19 July 2001). At the end of 2001, China became an official member of the World Trade Organization.
18. FAO, op. cit. note 11.

World's Rangelands Deteriorating Under Mounting Pressure
(pages 210–14)

1. "Desert Area Rises to 28 Percent," *Deutsche Presse-Agentur*, 29 January 2002.
2. Human population from United Nations, *World Population Prospects: The 2000 Revision* (New York: February 2001); livestock data from U.N. Food and Agriculture Organization (FAO), *FAOSTAT Statistics Database*, at <apps.fao.org>, updated 2 May 2001.
3. Number of pastoralists from "Investing in Pastoralism," *Agriculture Technology Notes* (Rural Development Department, World Bank), March 1998, p. 1; livestock numbers from FAO, op. cit. note 2; land area estimate from Stanley Wood, Kate Sebastian, and Sara J. Scherr, *Pilot Analysis of Global Ecosystems: Agroecosystems* (Washington, DC: International Food Policy Research Institute and World Resources Institiute, 2000), p. 3.

4. Land from Wood, Sebastian, and Scherr, op. cit. note 3.
5. Africa's 3 million buffalo are included in the estimate for cattle, found in FAO, op. cit. note 2; Southern African Development Coordination Conference, *SADCC Agriculture: Toward 2000* (Rome: FAO, 1984).
6. Edward C. Wolf, "Managing Rangelands," in Lester Brown et al., *State of the World 1986* (New York: W.W. Norton & Company, 1986); Government of India, "Strategies, Structures, Policies: National Wastelands Development Board," New Delhi, mimeographed, 6 February 1986.
7. FAO, op. cit. note 2.
8. Erik Eckholm, "Chinese Farmers See a New Desert Erode Their Way of Life," *New York Times*, 30 July 2000.
9. FAO, op. cit. note 2; United Nations, op. cit. note 2.
10. H. Dregne et al., "A New Assessment of the World Status of Desertification," *Desertification Control Bulletin*, no. 20, 1991, cited in Lester R. Brown and Hal Kane, *Full House* (New York: W.W. Norton & Company, 1994), p. 95.
11. Ibid.; gross domestic product from International Monetary Fund, *World Economic Outlook (WEO) Database*, at <www.imf.org/external/pubs/ft/weo/2000/02/data/index.htm>, September 2000.
12. Central Budget and Planning Organization, and Statistics and Registration Administration of Iran, cited by Farzaneh Bahar, Former General Director of Family Planning in Iran's Khorasan state, 23 December 2001, e-mail to Janet Larsen, Earth Policy Institute; Abubakar Dungus, "Iran's Other Revolution," *Populi*, September 2000.
13. FAO, op. cit. note 2; A. Banerjee, "Dairying Systems in India," *World Animal Review*, vol. 79/2 (Rome: FAO, 1994); S. C. Dhall and Meena Dhall, "Dairy Industry—India's Strength Is in Its Livestock," *Business Line*, Internet Edition of *Financial Daily* from *The Hindu* group of publications, at <www.indiaserver.com/businessline/1997/11/07/stories/ 03070311.htm>, 7 November 1997.
14. China's crop residue production and use from Gao Tengyun,

"Treatment and Utilization of Crop Straw and Stover in China," *Livestock Research for Rural Development*, February 2000; China's "Beef Belt" from U.S. Department of Agriculture, Economic Research Service, "China's Beef Economy: Production, Marketing, Consumption, and Foreign Trade," *International Agriculture and Trade Reports: China* (Washington, DC: July 1998), p. 28.

15. See <www.icarda.cgiar.org>.

Fish Farming May Overtake Cattle Ranching As a Food Source
(pages 215–220)

1. U.N. Food and Agricultural Organization (FAO), *Yearbook of Fishery Statistics: Capture Production* and *Aquaculture Production* (Rome: various years).
2. In Figure 3–3, fish catch from ibid. and beef production from FAO, *1948–1985 World Crop and Livestock Statistics* (Rome: 1987); FAO, *FAOSTAT Statistics Database*, at <apps.fao.org>, updated 2 May 2001.
3. Conversion ratio for grain to beef based on Allen Baker, Feed Situation and Outlook staff, U.S Department of Agriculture (USDA), Economic Research Service (ERS), Washington, DC, discussion with author, 27 April 1992; conversion ratio for fish from USDA, ERS, "China's Aquatic Products Economy: Production, Marketing, Consumption, and Foreign Trade," *International Agriculture and Trade Reports: China* (Washington, DC: July 1998), p. 45.
4. FAO, *Yearbook of Fishery Statistics: Aquaculture Production 1998*, vol. 86/2 (Rome: 2000).
5. Ibid.
6. Ibid; population from United Nations, *World Population Prospects: The 2000 Revision* (New York: February 2001).
7. FAO, op. cit. note 4.
8. K. J. Rana, "China," in *Review of the State of World Aquaculture*, FAO Fisheries Circular No. 886 (Rome: 1997), at <www.fao.org/fi/publ/circular/c886.1/c886-1.asp>; information on rice and fish polyculture from Li Kangmin, "Rice

Aquaculture Systems in China: A Case of Rice-Fish Farming from Protein Crops to Cash Crops," *Proceedings of the Internet Conference on Integrated Biosystems 1998* at <www.ias.unu.edu/proceedings/icibs/li/paper.htm>, viewed 5 July 2000.

9. Information on China's carp polyculture from Rosamond L. Naylor et al., "Effect of Aquaculture on World Fish Supplies," *Nature*, 29 June 2000, p. 1022; polyculture in India from W. C. Nandeesha et al., "Breeding of Carp with Oviprim," in Indian Branch, Asian Fisheries Society, *Special Publication No. 4* (Mangalore, India: 1990), p. 1.

10. Krishen Rana, "Changing Scenarios in Aquaculture Development in China," *FAO Aquaculture Newsletter*, August 1999, p. 18.

11. Catfish feed requirements from Naylor et al., op. cit. note 9, p. 1019; U.S. catfish production data from USDA, ERS, National Agriculture Statistics Service, *Catfish Production* (Washington, DC: July 2000), p. 3.

12. FAO, op. cit. note 4.

13. Naylor et al., op. cit. note 9.

14. Rosamond L. Naylor et al., "Nature's Subsidies to Shrimp and Salmon Farming," *Science*, 30 October 1998, pp. 883–84.

15. Ibid.

16. Population from United Nations; op. cit. note 6.

Our Closest Relatives Are Disappearing (pages 221–25)

1. John F. Oates et al., "Extinction of a West African Red Colobus Monkey," *Conservation Biology*, 5 October 2000, pp. 1526–32.

2. Species Survival Commission (SSC), Craig Hilton-Taylor, compiler, *2000 IUCN Red List of Threatened Species* (Gland, Switzerland, and Cambridge, U.K.: World Conservation Union–IUCN, 2000).

3. Ibid.

4. John Tuxill, "Death in the Family Tree," *World Watch*, September/October 1997, p. 14.
5. Human population from United Nations, *World Population Prospects: The 2000 Revision* (New York: February 2001).
6. Russell Mittermeier, "Biodiversity Issues Affecting Primates," keynote paper from *The Apes: Challenges for the 21st Century*, Brookfield Zoo, Chicago, May 2000; habitat loss from SSC, op. cit. note 2.
7. Deforestation and fires from Forest Watch Indonesia and Global Forest Watch, *The State of the Forest: Indonesia* (Bogor, Indonesia, and Washington, DC: 2002), p. xi; orangutan numbers from Carel P. van Schaik, "Securing a Future for the Wild Orangutan," keynote paper from *The Apes: Challenges for the 21st Century*, op. cit. note 6.
8. Danna Harman, "Bonobos' Threat to Hungry Humans," *Christian Science Monitor*, 7 June 2001.
9. Kari Lydersen, "Congo War Threatens a Pacifist Human Relative," *Washington Post*, 5 June 2000; Gay Reinhartz and Inogwabini Bila Isia, "Bonobo Survival and a Wartime Conservation Mandate," keynote paper from *The Apes: Challenges for the 21st Century*, op. cit. note 6.
10. SSC, op. cit. note 2.
11. Bushmeat Crisis Task Force (BCTF), *Eco-Economics Fact Sheet* (Silver Spring, MD: April 2000); David S. Wilkie and Julia F. Carpenter, "Bushmeat Hunting in the Congo Basin: An Assessment of Impacts and Options for Mitigation," paper from *The Apes: Challenges for the 21st Century*, op. cit. note 6.
12. BCTF, op. cit. note 11; sustainable yield from John Robinson, Wildlife Conservation Society, speech at Bushmeat Crisis Task Force Capitol Hill Event, 18 May 2000.
13. Bushmeat consumption from BCTF, op. cit. note 11; African ape extinction from Jane Goodall, speech at Bushmeat Crisis Task Force Capitol Hill Event, op. cit. note 12.
14. Convention on International Trade in Endangered Species of Wild Fauna and Flora available at <www.cites.org>.

Illegal Logging Threatens Ecological and Economic Stability (pages 226–30)

1. "Jakarta Floods Uncover System Faults: Illegal Logging, Judicial Bias Compound Indonesia's Woes," *Nikkei Weekly* (Japan), 18 February 2002.

2. Forest Watch Indonesia and Global Forest Watch, *The State of the Forest: Indonesia* (Bogor, Indonesia, and Washington, DC: 2002), pp. xi, 3.

3. Ibid., pp. xi, 36.

4. Michael Bengwayan, "Illegal Logging Wipes Out Philippine Forests," *Environment News Service*, 11 October 1999; Chris Brown, Patrick B. Durst, and Thomas Enters, *Forests Out of Bounds: Impacts and Effectiveness of Logging Bans in Natural Forests in Asia-Pacific* (Bangkok: U.N. Food and Agriculture Organization, Asia-Pacific Forestry Commission, October 2001).

5. Thailand from Brown, Durst, and Enters, op. cit. note 4; Yangtze river basin deforestation from Carmen Revenga et al., *Watersheds of the World* (Washington, DC: World Resources Institute and Worldwatch Institute, 1998); John Pomfret, "China's Lumbering Economy Ravages Border Forests," *Washington Post*, 26 March 2001.

6. China's timber balance from Sun Xiufang and Ralph Bean, *China: Solid Wood Products Annual, 2001* (Beijing: U.S. Department of Agriculture (USDA), Foreign Agricultural Service (FAS), Global Agriculture Information Network (GAIN) Report, 18 July 2001); International Timber Trade Organization projection from Pomfret, op. cit. note 5; Japan from Fred Pearce, "Logging Ban Backfires," *New Scientist*, 3 March 2001.

7. Sun and Bean, op. cit. note 6; "G-7 Nations and China Must Halt the Import of Illegal Timber from the Russian Far East," press release (Gland, Switzerland: World Wide Fund for Nature, 27 February 2002).

8. Pomfret, op. cit. note 5.

9. Laos and Viet Nam from "The Fight Against Illegal Loggers,"

The Economist, 3 April 1999, and from Dave Currey et al., *Timber Trafficking: Illegal Logging in Indonesia, South East Asia and International Consumption of Illegally Sourced Timber* (London: Emerson Press, Environmental Investigation Agency and Telapak Indonesia, September 2001), pp. 9–10; Cambodia from Jim Ford and Alexander Sheingauz, "Major Trends and Issues in Forests and Forestry: Globally and in Russia," Conference on Commercial Forestry in the Russian Far East: Opportunities for Sustainable Trade, Conservation and Community Development (Yuzhno-Sakhalinsk, Russia: Forest Trends, Economic Research Institute, 18–20 September 2001).

10. Susan Minnemeyer, *An Analysis of Access Into Central Africa's Rainforests* (Washington, DC: Global Forest Watch and World Resources Institute, 2002), p. 11.

11. Ibid.; Mark Jaffe, "Logging Fuels Crime, Corruption in Cameroon," *Philadelphia Inquirer*, 21 May 2001; Minnemeyer, op. cit. note 10; Ford and Sheingauz, op. cit. note 9.

12. Brazil from Currey et al., op. cit. note 9; Mexico from Nick Miles, "Mexico's 'Devastating' Forest Loss," *BBC News*, 4 March 2002; Ethiopia from Dechassa Lemessa and Matthew Perault, *Forest Fires in Ethiopia: Reflections on Socio-Economic and Environmental Effects of the Fires in 2000* (Addis Ababa: United Nations Development Programme–Emergencies Unit for Ethiopia, 7 December 2001), p. 1.

13. Currey et al., op. cit. note 9, p. 4.

14. Michael Smith and Mikhail Maximenko, *Russian Federation Solid Wood Products Annual, 2002* (Moscow: USDA, FAS, GAIN Report, 22 February 2002), p. 5.

Green Power Purchases Growing (pages 231–35)

1. City of Chicago, Office of the Mayor, "City Selects ComEd to Provide Clean Power," press release (Chicago: 6 June 2001); Green Power Network, "States with Competitive Green Power Offerings," at <www.eren.doe.gov/greenpower/dereg_map.html>, updated May 2001.

2. "Making the Switch: Why Britain's Universities Must Switch to Green Electricity," *People & Planet Online*, at <www.

peopleandplanet.org/climatechange/switch.asp#no7>, viewed 14 February 2002.

3. "Number of Green Energy Customers in The Netherlands Tripled in 2001," *Greenprices*, 28 January 2002, at <www.greenprices.com/nl/newsitem.asp?nid=283>.

4. "Greenprices: Green Energy in Germany," at <www.greenprices.com/de/index.asp>, viewed 15 February 2002, and at <www.greenprices.com/de/usertop.asp>, viewed 20 February 2002.

5. "Greenprices: Green Energy in Germany—Policy on Renewable Energy," at <www.greenprices.com/de/greenpol.asp>, viewed 20 February 2002.

6. "Green Power: Frequently Asked Questions," at <www.greenpower.com.au/GPFaq.shtml#GP7>, viewed 20 February 2002; sources of green energy from "Approved Green Power Generators—January 2002," at <www.greenpower.com.au/images/dl/GPGensJan02.pdf>, viewed 20 February 2002.

7. Grassroots Campaign for Wind Power, at <www.cogreenpower.org/Parade.htm>, updated December 2000; "University of Colorado Students Vote 'Yes' for Wind Power!" press release (Boulder, CO: 17 April 2000).

8. Green Power Network, "U.S. Green Marketing Activities: Customers," at <www.eren.doe.gov/greenpower/mkt_customer. html>, viewed 28 February 2002.

9. Blair Swezey and Lori Bird, "Businesses Lead the 'Green Power' Charge," *Solar Today*, January/February 2001, p. 24.

10. Environmental Protection Agency, *Green Power: Fueling EPA's Mission With Renewable Energy* (Washington, DC: December 2001), p. 5; Green Power Network, op. cit. note 7.

11. Center for Resource Solutions, "Green-e Standard," at <www.green-e.org/ipp/standard_for_marketers.html>, revised 16 August 2001; Gabe Petlin, Center for Resource Solutions, discussion with author, 7 March 2002.

12. "Illinois Initiates Green Power Standard," *SolarAccess.com Daily News*, 25 March 2002.

New York: Garbage Capital of the World (pages 236–40)

1. Kirk Johnson, "The Mayor's Budget Proposal: Recycling; Glass, Metal and Plastic May Become Plain Trash," *New York Times*, 14 February 2002; Kirk Johnson, "To City's Burden, Add 11,000 Tons of Daily Trash," *New York Times*, 28 February 2002.
2. Eric Lipton, "The Long and Winding Road Now Followed by New York City's Trash," *New York Times*, 24 March 2001.
3. Calculations by author; Lhota quoted in Lipton, op. cit. note 2.
4. Gilmore quoted in Lipton, op. cit. note 2.
5. "Virginia Gov. Proposes Plan to Add Solid Waste Fee," *Reuters*, 10 April 2002.
6. Johnson, "To City's Burden," op. cit. note 1.
7. Columbia University's Earth Institute, *Life After Fresh Kills: Moving Beyond New York City's Current Waste Management Plan* (New York: Earth Engineering Center and Urban Habitat Project, 1 December 2001), p. B-31.
8. Kirk Johnson, "As Options Shrink, New York Revisits Idea of Incineration," *New York Times*, 23 March 2002.
9. Columbia University's Earth Institute, op. cit. note 7, p. A-12.
10. International examples from Brenda Platt and Neil Seldman, *Wasting and Recycling in the United States 2000* (Athens, GA: GrassRoots Recycling Network, 2000).
11. Janet N. Abramovitz and Ashley T. Mattoon, *Paper Cuts: Recovering the Paper Landscape*, Worldwatch Paper 149 (Washington, DC: Worldwatch Institute, December 1999), p. 39.
12. U.S. Department of the Interior, U.S. Geological Survey, *Mineral Commodity Summaries 2001* (Washington, DC: 2001).

Tax Shifting on the Rise (pages 241–45)

1. David Roodman, "Environmental Tax Shifts Multiplying," in

Lester R. Brown et al., *Vital Signs 2000* (New York: W.W. Norton & Company, 2000), pp. 138–39.

2. J. Andrew Hoerner and Benoît Bosquet, *Environmental Tax Reform: The European Experience* (Washington, DC: Center for a Sustainable Economy, February 2001), pp. 17–18.
3. Figure of 2 percent from Kai Schlegelmilch, German Ministry of the Environment, e-mail to author, 2 June 2002; fuel sales, gas consumption, and carpool growth from German Ministry of the Environment, "Environmental Effects of the Ecological Tax Reform," at <www.bmu.de/english/topics/oekosteuer/oekosteuer_environment.php>, viewed 20 May 2002.
4. Hoerner and Bosquet, op. cit. note 2; European Environmental Bureau Campaign on Environmental Fiscal Reform—Germany, at <www.ecotax.info/germany.htm>, viewed 17 May 2002.
5. European Environment Agency (EEA), *Environmental Taxes: Recent Developments in Tools for Integration* (Copenhagen: November 2000), pp. 75–77; complete conversion to ultra-low sulfur diesel from Andrew Field, U.K. Treasury, e-mail to author, 24 June 2002.
6. EEA, op. cit. note 5, pp. 68–69; Hoerner and Bosquet, op. cit. note 2, p. 20.
7. European Environmental Bureau Campaign on Environmental Fiscal Reform—Netherlands, at <www.ecotax.info/netherlands.htm>, viewed 20 May 2002.
8. EEA, op. cit. note 5, p. 65.
9. European Environmental Bureau Campaign on Environmental Fiscal Reform—Sweden, at <www.ecotax.info/sweden.htm>, viewed 20 May 2002; "Sweden Makes Major Reductions in GHGs," *Xinhua News Agency*, 16 January 2002.
10. Organisation for Economic Co-operation and Development (OECD), *Environmentally Related Taxes in OECD Countries: Issues and Strategies* (Paris: 2001), p. 124.
11. David Malin Roodman, *Paying the Piper: Subsidies, Politics, and the Environment*, Worldwatch Paper 133 (Washington,